Springer Undergraduate Mathematics Series

Advisory Editors

M. A. J. Chaplain, St. Andrews, UK
Angus Macintyre, Edinburgh, UK
Simon Scott, London, UK
Nicole Snashall, Leicester, UK
Endre Süli, Oxford, UK
Michael R. Tehranchi, Cambridge, UK
John F. Toland, Bath, UK

The Springer Undergraduate Mathematics Series (SUMS) is a series designed for undergraduates in mathematics and the sciences worldwide. From core foundational material to final year topics, SUMS books take a fresh and modern approach. Textual explanations are supported by a wealth of examples, problems and fully-worked solutions, with particular attention paid to universal areas of difficulty. These practical and concise texts are designed for a one- or two-semester course but the self-study approach makes them ideal for independent use.

More information about this series at http://www.springer.com/series/3423

Jonathan Gillard

A First Course in Statistical Inference

 Springer

Jonathan Gillard
School of Mathematics
Cardiff University
Cardiff, UK

ISSN 1615-2085   ISSN 2197-4144 (electronic)
Springer Undergraduate Mathematics Series
ISBN 978-3-030-39560-5   ISBN 978-3-030-39561-2 (eBook)
https://doi.org/10.1007/978-3-030-39561-2

Mathematics Subject Classification (2010): 62-01, 62D05, 62E15, 62F03, 62F10, 62J05, 62J10

This Springer imprint is published by the registered company Springer Nature Switzerland AG
The registered company address is: Gewerbestrasse 11, 6330 Cham, Switzerland

This book is dedicated to three special people: my wife Becky, my son Benji and my daughter Rosie. It is written in loving memory of my mother Shirley. I am forever grateful for the unconditional love.

Preface

Context

This book is the result of efforts by the author to teach introductory statistical inference to first-year mathematics undergraduate students over the past 10 years. When beginning to develop my introductory course, I struggled to find an appropriate textbook that I could recommend to students. Many were too weighty, aiming to give the full coverage of statistical methodology. Others, particularly those with a mathematics focus, were too advanced for students just entering university and possibly encountering statistics as part of their mathematics degree for the first time.

Statistics is a challenging topic for students. The prose in this book is largely conversational to help explain to students the reasoning and logic behind certain concepts. This is balanced with a degree of mathematical formalism which should be familiar to mathematics undergraduates. Key theorems are stated and proven, and important definitions are provided.

Aim

This book aims to fill the gap described above. It contains sufficient material for a total of 25–30 h worth of lecturing, specifically designed for mathematics students encountering statistical inference for the first time. It is assumed that the student is familiar with some probability. They should be competent with elementary probability operations, for example, being able to compute $P[X < 4]$ where X is a Normally distributed random variable. Some knowledge of the Binomial, Poisson, Exponential, and Normal distributions would be helpful, as would competence with calculating the expected value and variance of a linear combination of independent random variables. These topics are visited and explained in some detail in this book (particularly in Chap. 1), but someone not familiar with these topics may need a gentler introduction.

The book strikes a balance between developing the mathematical principles of the core ideas of inference, namely, estimation, confidence intervals, and hypothesis testing (which underpin most of statistical practice) and giving students the opportunity to start to develop their statistical intuition. The examples used aim to demonstrate the real-life applicability of the subject. Exercises are also provided, along with full solutions at the rear of the book.

This book can be used as a starting point to cover the key concepts of statistics. The book ends with describing ANOVA and how to fit a straight line to paired data. These two topics are included for two purposes: to reinforce the core ideas of confidence intervals and hypothesis testing, and to describe two common practical uses of statistics.

Use of R

To help illuminate some of the ideas, examples and experiments using R are provided. This isn't a textbook on the specifics or use of R. For those not familiar with R, these clearly highlighted sections may be ignored. R is a free software environment for statistical computing and graphics. Information and where to download it is available at https://www.r-project.org/. A brief introduction to R is provided at the appendix of the book.

Cardiff, UK   Jonathan Gillard
December 2019

Contents

Recap of Probability Fundamentals

1.1 Introduction

The aim of this chapter is to give a brief recap of the key probability ideas, tools, and techniques that will be used in subsequent chapters. Specifically, we will describe

- Random variables (both discrete and continuous),
- Probability mass/density functions, and
- How to compute the mean and variance of random variables.

If you are familiar with these concepts, then you can jump to the next chapter. The aim here is not to give a comprehensive treatment of these topics but to serve as a recap.

1.2 Random Variables

1.2.1 Discrete and Continuous Random Variables

Let's toss a coin three times. From each toss, we can obtain a head (H) or a tail (T). The set of all possible outcomes is given by

$$\{HHH, HHT, HTH, THH, HTT, THT, TTH, TTT\}. \tag{1.1}$$

We will assume that each toss of the coin is independent. Generally speaking, we say events are independent if the occurrence of one does not affect the probability of occurrence of the other. This assumption seems reasonable in this scenario and means that the probability of observing HHH can be found by the product of the

J. Gillard, *A First Course in Statistical Inference*,
Springer Undergraduate Mathematics Series,
https://doi.org/10.1007/978-3-030-39561-2_1

probabilities of achieving a H on each toss. Indeed, each of these listed outcomes is equally likely with probability $\frac{1}{2} \times \frac{1}{2} \times \frac{1}{2} = \frac{1}{8}$.

Suppose that we are solely interested in the number of heads obtained, and denote this by X. Here X is a so-called *random variable*. It is a variable whose values depend on outcomes of a random phenomenon. There are many other examples of random variables, such as the amount of rain collected in a cup placed outside for a month; the number of visitors to a fast food restaurant in an hour; the heights of students attending a particular lecture. Random variables are revisited in Chap. 2. They are conventionally denoted by capital letters, and the values they take are denoted by lower case letters.

For our example of tossing three coins simultaneously, we are able to describe all possible values the random variable X can take and consider the probability of obtaining each of the outcomes. By inspecting the set of all possible outcomes (1.1), the random variable X can take the values 0, 1, 2, or 3, and we can deduce the probability of obtaining each one of these outcomes. To obtain two heads, i.e., to observe that $X = 2$, then we have to observe one of the following outcomes: $\{HHT, HTH, THH\}$. Each of these outcomes listed occurs with probability $\frac{1}{8}$ and so the probability of obtaining two heads when three coins are tossed simultaneously is $\frac{3}{8}$. In brief, we can write $P[X = 2] = \frac{3}{8}$.

We can produce the following table which summarizes the probabilities of observing particular values of the random variable X:

x	0	1	2	3
$P[X = x]$	1/8	3/8	3/8	1/8

The random variable X is an example of a discrete random variable: our possible outcomes are discrete. We can also construct continuous random variables. For example, let us return to a random variable already mentioned, that of the amount of rain collected in a cup placed outside for a month. We suppose that the amount of rain collected could take any value between 100 and 200 mm. It is not possible in this scenario to summarize the probabilities of observing a particular value of the rainfall collected in a table like the one above. This is because we could observe 100, 100.01, 159.98363827 mm, and infinitely many other measurements of rainfall. The scale of measurement of rainfall is continuous and not discrete like our example of tossing a coin. Random variables concerning continuous scales of measurement are known as continuous random variables.

Example

A biased coin, for which the probability of obtaining a head is 3/5 and tail is 2/5, is tossed three times. Let X denote the random variable of the number of heads minus the number of tails. Form a table summarizing the probabilities of obtaining each possible outcome of X.

Solution: We first write a table which contains all possible outcomes upon tossing the coin three times, the value of X for each of these outcomes along with the probabilities of each outcome happening. We again assume that each toss of the coin is independent, meaning that the probability of each outcome can be obtained by the appropriate product of the probability of observing H or T depending on the particular outcome.

Outcome	X	Probability
HHH	3	$(3/5)^3$
HHT	1	$(3/5)^2(2/5)$
HTH	1	$(3/5)^2(2/5)$
THH	1	$(3/5)^2(2/5)$
HTT	-1	$(3/5)(2/5)^2$
THT	-1	$(3/5)(2/5)^2$
TTH	-1	$(3/5)(2/5)^2$
TTT	-3	$(2/5)^3$

We now have all of the information needed to form a table summarizing the probabilities of obtaining each possible outcome of X. This is given below:

x	-3	-1	1	3
$P[X = x]$	8/125	36/125	54/125	27/125

Here, for example, we are able to calculate $P[X = 1]$ by recognizing that the event $X = 1$ happens upon observing one of the following events: HHT, HTH, or THH. Any of these individual events happens with probability $(3/5)^2(2/5) = 18/125$ and so $P[X = 1] = 18/125 + 18/125 + 18/125 = 54/125$.
◀

In the case of having a continuous random variable, we cannot tabulate the probabilities of obtaining outcomes as above. In this case, we can try to describe the probabilities of obtaining certain outcomes by a function. Indeed, one could imagine scenarios of discrete random variables where presenting the probabilities in a table as above rather cumbersome. The next section will describe situations where the probabilities of obtaining outcomes for discrete and continuous random variables can be summarized by a function. Such functions are called probability mass functions for discrete random variables and probability density functions for continuous random variables.

1.2.2 Probability Mass/Density Functions

Probability mass/density functions are used to calculate the probability of obtaining particular outcomes. We will take discrete and continuous random variables in turn.

1.2.2.1 Discrete Random Variables and Probability Mass Functions
We'll begin with an example. Let X be the discrete random variable denoting the number of tosses of a fair coin until a head is obtained. Assuming that consecutive

tosses are independent, then we can summarize the probability of obtaining particular outcomes in the following table:

x	1	2	3	4	$\cdots$
$P[X = x]$	1/2	1/4	1/8	1/16	$\cdots$

This is an example of a discrete random variable whose information about obtaining particular outcomes is not ideally displayed in a table because there are infinitely many possible outcomes. We do have a concise form we can use to describe the probability of obtaining a particular outcome. By inspection we can deduce that

$$P[X = x] = \left(\frac{1}{2}\right)^x, \quad \text{for } x = 1, 2, 3, 4, \ldots$$

Such a function is known as a probability mass function. Often, for conciseness, we let $f(x) = P[X = x]$ so we can use $f(x)$ as shorthand for the probability our random variable X takes the specific outcome x. The probability mass function $f(x)$ of a discrete random variable X has the following properties:

(a) $f(x) \geq 0$ for all x,
(b) $\sum_x f(x) = 1$, where the summation is over all possible values of x.

There is another function useful for describing a random variable, and this is the so-called cumulative distribution function. The cumulative distribution function of a random variable X is defined as $F(x) = P[X \leq x]$ for all x, which for a discrete random variable can be written as

$$F(x) = P[X \leq x] = \sum_{y \leq x} P[X = y].$$

Here the summation is over all possible values of y less than or equal to x. For the example of tossing three coins simultaneously, we can compute, for example,

$$F(3) = P[X \leq 3] = P[X = 1] + P[X = 2] + P[X = 3] = f(1) + f(2) + f(3)$$
$$= \frac{1}{2} + \frac{1}{4} + \frac{1}{8}$$
$$= \frac{7}{8}.$$

There are another two natural questions for this example. How many tosses of the coin should we expect to have to make before obtaining a head, and how might the number of tosses vary? These questions can be answered by considering the mean (or expectancy) and variance of the random variable X.

We define the mean and variance of a discrete random variable to be

$$E[X] = \sum_x x P[X = x] \tag{1.2}$$

$$Var[X] = E[X^2] - (E[X])^2 = \sum_x x^2 P[X = x] - (E[X])^2 \tag{1.3}$$

We demonstrate how to evaluate these quantities in the next example.

Example

For an online store, 10% of orders are delivered the following day, 50% take 2 days until they arrive, 30% take 3 days, and the rest take 4 days. Find the mean and variance of the delivery time of an order.

Solution: Let X be the random variable denoting delivery time in days. We can summarize the probabilities of each outcome from the information given as follows:

x	1	2	3	4
$P[X = x]$	0.1	0.5	0.3	0.1

It follows that using (1.2)

$$E[X] = (1 \times 0.1) + (2 \times 0.5) + (3 \times 0.3) + (4 \times 0.1) = 2.4,$$

and using (1.3)

$$Var[X] = E[X^2] - (E[X])^2$$
$$= (1 \times 0.1) + (4 \times 0.5) + (9 \times 0.3) + (16 \times 0.1) - (2.4)^2 = 0.64.$$

◀

Some particular cases of discrete random variables are important and are commonly used. Here is a list of two "famous" discrete random variables and their means and variances.

Binomial Distribution
Let us suppose that the result of an experiment can be classified into one of two possible categories. Examples include tossing a coin which results in a "head" or "tail", someone throwing a dart at a target can have a "success" or a "failure". One such experiment is known as a Bernoulli trial.

If we tossed a coin n times, then we have a sequence of n Bernoulli trials. If the trials are performed such that (i) all trials are independent and (ii) the probability of obtaining a "head" (or successfully hitting a target on a dartboard) is fixed and given by p, so the probability of obtaining a "tail" or failing to hit a target on a dartboard is $1 - p$ then we obtain something known as the Binomial distribution.

Let X be a random variable with Binomial distribution where there are n trials and probability of success p. As shorthand we typically write $X \sim Bin(n, p)$. It is possible to derive the following characteristics of the Binomial distribution:

Probability mass function: $f(x) = P[X = x] = \binom{n}{x} p^x (1 - p)^{n-x}$

Mean $E[X] = np$

Variance $Var[X] = np(1 - p)$

Example

Suppose a newborn baby is equally likely to be a girl or a boy. What is the probability that in five families each with two children, at most two families have two girls?

Solution: Assuming that the sexes of the children are independent, then the probability that a given family has two girls is $1/2 \times 1/2 = 1/4$. We can use the Binomial distribution to answer this question by labeling two girls as a "success" with probability $p = 1/4$ and all other outcomes as a "failure" with probability $3/4$.

Let X be the Binomial random variable of the number of families with two girls. We consider the five families as $n = 5$ independent trials. Then from the argument in the previous paragraph $X \sim Bin(5, 1/4)$, and by using the probability mass function, we can calculate the desired probability as

$$P[X \leq 2] = P[X = 0] + P[X = 1] + P[X = 2]$$
$$= \binom{5}{0}(1/4)^0(3/4)^5 + \binom{5}{1}(1/4)^1(3/4)^4 + \binom{5}{2}(1/4)^2(3/4)^3$$
$$= 459/512 \approx 0.8965 .$$

◄

Poisson Distribution

A distribution often appropriate for modeling the number of times an event occurs in a fixed time interval is known as the Poisson distribution. It is apt for describing events such as (i) the number of patients arriving at a hospital ward between 9 am and 11 am; (ii) the amount of mail received by an individual on a given day; and (iii) the number of decay events per minute from a radioactive source.

The Poisson distribution is typically a good approximation of such scenarios when the following assumptions hold:

- The number of times an event occurs, denoted by x, in a given time interval, takes the values $0, 1, 2, \ldots$
- Events occur independently.
- The average rate which events occur, denoted by λ, is constant. The parameter λ is known as the event rate.
- Two or more events cannot occur at precisely the same instant.

Let X be a random variable with Poisson distribution with event rate λ. As shorthand we typically write $X \sim Po(\lambda)$. It is possible to derive the following characteristics of the Poisson distribution:

Probability mass function: $f(x) = P[X = x] = \exp(-\lambda)\dfrac{\lambda^x}{x!}$

Mean $\qquad\qquad\qquad\qquad\qquad E[X] = \lambda$

Variance $\qquad\qquad\qquad\qquad Var[X] = \lambda$

Example

On average Becky receives one piece of post a day. What is the probability that Becky will receive at least two pieces of post on a randomly chosen day?

Solution: Let X be a random variable with Poisson distribution with mean $\lambda = 1$, that is, $X \sim Po(1)$. Then

$$P[X \geq 2] = 1 - P[X \leq 1] = 1 - (P[X = 0] + P[X = 1])$$
$$= 1 - \left(\exp(-1)\frac{1^0}{0!} + \exp(-1)\frac{1^1}{1!} \right)$$
$$\approx 0.26424 \,,$$

where here we have used the probability mass function of the Poisson distribution.
◀

1.2.2.2 Continuous Random Variables and Probability Density Functions

We now turn our attention to continuous random variables, where a random variable X takes a value in some continuous range. An example would be the time taken for a light bulb to expire. We will say that X is a continuous random variable if there exists a function $f(x)$, called the probability density function, such that

(a) $f(x) \geq 0$, for all x,
(b) $\int_{-\infty}^{+\infty} f(x)\,dx = 1$,
(c) for any numbers a and b with $a \leq b$,

$$P[a \leq X \leq b] = \int_a^b f(x)\,dx \,.$$

Note that for continuous random variables it is convention to call $f(x)$ the probability density function instead of probability mass function which is reserved for discrete random variables.

Just like before, we can also write down the cumulative distribution function for a continuous random variable X. It is given by

$$F(x) = P[X \leq x] = \int_{-\infty}^x f(x)\,dx$$

and there is an explicit relationship between the probability density function and cumulative distribution function for a continuous random variable whereby

$$f(x) = \frac{d}{dx}F(x) \,.$$

The mean or expectation of X can be found by computing

$$E[X] = \int_{-\infty}^{+\infty} x f(x) \, dx$$

with its variance found from

$$Var[X] = E[X^2] - (E[X])^2 = \int_{-\infty}^{+\infty} x^2 f(x) \, dx - (E[X])^2 .$$

In the next chapter, we introduce four key continuous distributions which are used heavily throughout this book. These are the so-called Normal, Student's t-, chi-squared, and F-distributions. We won't repeat their introduction here. To illustrate the definitions given above, we will describe another important continuous distribution, known as the Exponential distribution.

Exponential Distribution

The continuous random variable X has the Exponential distribution if its probability density function is given by

$$f(x) = \begin{cases} \lambda \exp(-\lambda x), & x \geq 0 \\ 0, & \text{otherwise} \end{cases}$$

for some $\lambda > 0$. As shorthand we typically write $X \sim Exp(\lambda)$.

Typical scenarios which can be modeled by an Exponential distribution include the lifetime of a light bulb, and the time between calls to a call center. Let X be a random variable with Exponential distribution with parameter λ. It is possible to derive the following characteristics of the Exponential distribution:

Cumulative density function: $F(x) = P[X \leq x] = 1 - \exp(-\lambda x)$

Mean $E[X] = \dfrac{1}{\lambda}$

Variance $Var[X] = \dfrac{1}{\lambda^2}$

The cumulative distribution function can be derived by simple integration of the probability density function with suitable limits, while the mean and variance can be found by using integration by parts. Try to derive these properties for yourself.

Example

The average time between calls to an ambulance call center is 0.2 min. Find the probability that the duration between calls is at least 0.4 min.

Solution: Let X be the random variable such that $X \sim Exp(\lambda)$. As the mean is 0.2 then $\lambda = 1/0.2 = 5$. Hence, the probability that the duration between calls is at least 0.4 min is

$$P[X > 0.4] = 1 - P[X \leq 0.4] = 1 - F(0.4) = 1 - (1 - \exp(-5 \times 0.4)) \approx 0.13534,$$

where here we have used the cumulative distribution function $F(x)$ of the Exponential random variable X.

◄

1.3 Exercises

1. Suppose a discrete random variable X has probability mass function

$$f(x) = P[X = x] = \begin{cases} kx^2, & \text{if } x = 1, 2, 3, \\ 0, & \text{otherwise.} \end{cases}$$

First find the value of k and thus derive

a. the mean of X,
b. the variance of X.

2. Two fair dice are tossed and X is the discrete random variable of the larger of the two scores obtained (or either score if they are equal). Find the probability mass function of X and determine $E[X]$.
3. The discrete random variable X has probability mass function

$$f(x) = P[X = x] = \begin{cases} \alpha(1 + \beta x), & \text{if } x = 0, 1, 2, 3, \\ 0, & \text{otherwise.} \end{cases}$$

where α and β are constants. Find α and $E[X]$ in terms of β. For the case $E[X] = 2$, find numerical values of α and β.
4. A die is rolled six times. What is the probability exactly one six is obtained?
5. The random variable X follows the Binomial distribution with $E[X] = 1$ and $Var[X] = 0.5$. What is $P[X \leq 1]$?
6. On average there are seven emergency admissions to a certain hospital each Saturday night. What is the probability that there will be fewer than two emergency admissions next Saturday night?
7. The continuous random variable X has the probability density function

$$f(x) = \begin{cases} kx(1 - x), & \text{if } 0 \leq x \leq 1, \\ 0, & \text{otherwise.} \end{cases}$$

First find the value of k and thus derive

a. the cumulative distribution function of X,
b. the mean of X,
c. the variance of X.

8. A bulb has a life X which has an Exponential distribution with mean $1/4$ years. Find the probability that the bulb will last for at least 1 year.

Sampling and Sampling Distributions

2

2.1 Introduction

For the moment, consider a population as a collection of objects, such as people, families, cars, etc. A sample is a subcollection or part of the population. Populations are studied because they have some interesting property or characteristic that varies among different members of the population. Such a characteristic is called a variable, e.g., the monthly income of families, the fuel consumption of a car, etc. A variable identifies a property of interest and is the basis upon which values are associated with members of the population. Formal definitions of these terms are as follows:

Definition 2.1 A *variable* is a rule that associates a value with each member of the population.

Definition 2.2 A *population* is the collection of all values of the variable under study.

Definition 2.3 A *sample* is any subcollection of the population.

Definition 2.4 A *population parameter* is some (often unknown) numerical measure associated with the population as a whole.

Definition 2.5 Let X denote a random variable. A *random sample* from X is a set of independent random variables $X_1, X_2, \ldots, X_n$, each with the same distribution as X. An alternative statement of this is to say $X_1, X_2, \ldots, X_n$ are i.i.d. (independent and identically distributed) random variables from (the distribution of) X.

Definition 2.6 The values taken by $X_1, X_2, \ldots, X_n$ in an observed sample are denoted by $x_1, x_2, \ldots, x_n$ and are called the *sample values*.

© Springer Nature Switzerland AG 2020
J. Gillard, *A First Course in Statistical Inference*,
Springer Undergraduate Mathematics Series,
https://doi.org/10.1007/978-3-030-39561-2_2

Definition 2.7 A *statistic* is a function of $X_1, X_2, \ldots, X_n$ and is thus itself a random variable. It does not contain any unknown parameters.

Let us consider an example to illuminate some of the above abstract definitions.

Example

A piggy bank contains six coins: one 5p, two 10p, one 20p, and two 50p coins. Consider the selection of two coins from the piggy bank, obtained with replacement.

For this example, let us consider what we mean by *variable, population, sample, population parameter, random sample, sample values,* and *statistic*.

Our *variable* in this case is the numerical value associated with each coin of the population. For example, one member of the population has the numerical value 5p, another two have the numerical value 10p, and so on.

Our *population* can be represented by the set $\{5p, 10p, 10p, 20p, 50p, 50p\}$. In most real-life applications, the population is not as easily listed as this (because the population is too large, or inaccessible) and convention is to take a *sample* from it to try to understand the population. A sample of size two from this population could be $\{5p, 10p\}$, or $\{10p, 10p\}$, and so on. We'll revisit this in a moment.

Population parameters for our population could be the population mean, population variance, population minimum, and anything else that we could think of. It is any numerical measure of the population. Typically, we wouldn't know these parameters, but as our population of six coins is so small we could actually compute them. But let us consider a case where it might not be so easy. Suppose we consider the population of heights of commuters currently at London Paddington railway station. There is currently a population mean height, a variance of these heights as well as a minimum. It would be exceptionally time-consuming to measure everyone's heights to obtain these population parameters. But we could take a sample of say 20 people, measure their heights, and try to use this data to guess what the population parameters may be.

We could let X be the *random variable* representing a randomly chosen coin. This allows us to use some shorthand. Instead of saying take a random sample of size two from $\{5p, 10p, 10p, 20p, 50p, 50p\}$, we could say take a random sample of size two from X. By random sample, we mean that the probability of obtaining a particular coin is not affected by what came before it, and the probability distribution of picking a coin doesn't change whether it is the first or second pick. In this example, it would mean we are sampling with replacement. Let us take a random sample of size two. On my first pick I could obtain $5p$. I then return this coin to the piggy bank and on my second pick I could obtain $5p$ again.

If we have agreed that we are going to take a random sample of size two from X, it is useful to have some notation. We could denote this random sample by $\{X_1, X_2\}$ where X_1 would denote all possible first picks (anything from $\{5p, 10p, 10p, 20p, 50p, 50p\}$) and X_2 would denote all possible second picks (again anything from $\{5p, 10p, 10p, 20p, 50p, 50p\}$). Once we have made our picks, it is useful to have some notation to refer to them. This is usually small-case letters of the random variable from which they have come, e.g., x_1 and x_2. Note that X_1 and X_2 are random variables, but x_1 and x_2 are, in this example, scalars. Because of the way we have taken our sample, X_1 and X_2 are independent and have identical distributions.

Once we have our sample we can compute statistics. But we take samples randomly. So I could take a random sample of size two from $\{5p, 10p, 10p, 20p, 50p, 50p\}$, and compute the mean of these numbers. If I then took a different random sample of size two from $\{5p, 10p, 10p, 20p, 50p, 50p\}$, I'll likely get two different numbers, and thus obtain a different sample mean. A *statistic* is also a random variable because we need to take a sample at random in order to compute it in the first place. The value of the statistic depends on the sample, which, throughout this book, is generated randomly.

◀

We next consider some elementary statistics, which you may have seen before. Suppose we have taken a random sample $X_1, X_2, \ldots, X_n$ from a random variable X and have observed the numbers $x_1, x_2, \ldots, x_n$.

Definition 2.8 The *sample mean* of $x_1, x_2, \ldots, x_n$ is given by $\bar{x} = \dfrac{1}{n} \displaystyle\sum_{i=1}^{n} x_i$.

Definition 2.9 The *sample variance* of $x_1, x_2, \ldots, x_n$ is given by

$$s^2 = \frac{1}{n-1} \sum_{i=1}^{n} (x_i - \bar{x})^2 .$$

By expanding the brackets and simplifying, note that the sample variance s^2 may also be written as

$$s^2 = \frac{1}{n-1} \left\{ \sum_{i=1}^{n} x_i^2 - n\bar{x}^2 \right\} = \frac{1}{n-1} \left\{ \sum_{i=1}^{n} x_i^2 - \frac{\left(\sum_{i=1}^{n} x_i \right)^2}{n} \right\} .$$

You've no doubt encountered the sample mean before. What might look strange about the sample variance is that there is an $(n-1)$ on the denominator, and not an n. The reason for this will become evident later.

2.2 Sampling Distributions

This section will try to further motivate the use of taking a sample. The main idea is that we have a population that we wish to understand. But because of size, cost, or some other reasons, we are unable to see everyone in this population. Some populations are clearly too large, so even if we are interested in the distribution of heights of Europeans, for example, measuring every European would be a significant undertaking. Conversely, if we are interested in alcoholism, alcoholics may not be willing to readily identify themselves and thus identifying this population would be difficult. Instead, we take a sample (because it is cheaper, or quicker, or the only thing possible) and use this sample to make sensible guesses about the population. We will revisit the example of the previous section to see how this could work.

Example

A piggy bank contains six coins: one 5p, two 10p, one 20p, and two 50p coins. Enumerate all possible random samples of size two from the piggy bank, the probabilities of obtaining each of these random samples, and enumerate all possible sample means and variances. Consider how these sample means and variances relate to the population mean and variance.

Solution: Let X be the random variable representing the population. We can construct the probability distribution of X:

x	5p	10p	20p	50p
$P[X = x]$	1/6	2/6	1/6	2/6

and we can compute the mean and variance of this population. We have, for example, that

$$E[X] = (5 \times 1/6) + (10 \times 2/6) + \cdots (50 \times 2/6) = \frac{24}{6}$$

using (1.2) and

$$Var[X] = E[X^2] - (E[X])^2$$

$$= (5^2 \times 1/6) + (10^2 \times 2/6) + \cdots (50^2 \times 2/6) - \left(\frac{24}{6}\right)^2 \approx 353.47$$

using (1.3). Typically, these would be population parameters which we wouldn't know, but for this toy example of a known population we can compute them. We can now try to see how the sample means and sample variances relate to the equivalent population parameters.

To enumerate all possible random samples (along with their sample means and variances) and the probabilities of obtaining each sample, it is useful to produce a table such as that below:

(x_1, x_2)	No. of ways	$P[(X_1, X_2) = (x_1, x_2)]$	$\bar{x}$	s^2
(5p,5p)	1	1/36	5	0
(5p,10p)	4	4/36	7.5	12.5
(5p,20p)	2	2/36	12.5	112.5
(5p,50p)	4	4/36	27.5	1012.5
(10p,10p)	4	4/36	10	0
(10p,20p)	4	4/36	15	50
(10p,50p)	8	8/36	30	800
(20p,20p)	1	1/36	20	0
(20p,50p)	4	4/36	35	450
(50p,50p)	4	4/36	50	0

The first column can be seen to contain all possible random samples of size two. There are a number of ways of computing the probability of a particular sample occurring. Because our definition of a random sample coincides with sampling with replacement, there are $6 \times 6 = 36$ possible configurations of samples of size two. We have to choose one of 6 coins for our first random pick, and one of 6 coins again for our second random pick. There is only one possible way of obtaining a sample of two $5p$ coins: we must select the one $5p$ coin on our first pick, and then select it again on our second pick. So the probability of us getting the sample $(5p, 5p)$ is $1/6 \times 1/6 = 1/36$. A more difficult example is given by the next row. A sample of $(5p, 10p)$ could happen in a number of ways. Let us label one of our $10p$ coins as $10p(a)$ and another as $10p(b)$. Then, we could get a configuration of $(5p, 10p(a))$, or $(5p, 10p(b))$. We could, instead, get one of the $10p$ coins out on the first pick. So there are four different configurations which give rise to a sample of $(5p, 10p)$, and so the probability of obtaining such a sample is $4/36$. The final two columns for the sample mean and sample variance are computed using formulae from Definitions 2.8 and 2.9.

The above table lists all possible sample means and all possible sample variances that we could obtain, with the probability of achieving them readily deduced. For example, obtaining a sample variance of 0 is only possible if we observe one of the following samples: $(5p, 5p)$ or $(10p, 10p)$ or $(20p, 20p)$ or $(50p, 50p)$. The probability of getting any of these samples is $1/36$, $4/36$, $1/36$, and $4/36$, respectively, and so the probability of getting a sample variance of 0 is $1/36 + 4/36 + 1/36 + 4/36 = 10/36$. We can repeat such calculations and make probability distributions for each sample statistic. A probability distribution for a sample statistic is known as a *sampling distribution*.

The sampling distribution for the sample mean $\bar{X}$ is

$\bar{x}$	5	7.5	10	12.5	15	20	27.5	30	35	50
$P[\bar{X} = \bar{x}]$	1/36	4/36	4/36	2/36	4/36	1/36	4/36	8/36	4/36	4/36

and the sampling distribution for the sample variance S^2 is

s^2	0	12.5	50	112.5	450	800	1012.5
$P[S^2 = s^2]$	10/36	4/36	4/36	2/36	4/36	8/36	4/36

Note that we use capital letters for the random variables corresponding to the sample mean and sample variance: this is to explicitly acknowledge that they are based on a random sample, and so are random variables in themselves.

Now we have these sampling distributions (which are nothing more than probability distributions), we make the following calculations:

$$E[\bar{X}] = (5 \times 1/36) + (7.5 \times 4/36) + \cdots + (50 \times 4/36) = \frac{24}{6} = E[X],$$

$$Var[\bar{X}] = E[\bar{X}^2] - \left(E[\bar{X}]\right)^2 = (5^2 \times 1/36) + \cdots + (50^2 \times 4/36) - \left(\frac{24}{6}\right)^2$$

$$= 176.736 = \frac{Var[X]}{2},$$

and $E[S^2] = (0 \times 10/36) + (12.5 \times 4/36) + \cdots + (1012.5 \times 4/36) = 353.47 = Var[X]$. We can also compute $Var[S^2]$ if we wish. We can also repeat the same exercise for other sample statistics, such as the sample minimum or sample maximum.

These results are interesting. We have demonstrated that we should expect the sample mean to equal the population mean, and expect the sample variance to equal the population variance. We can also see that how much our sample mean varies is related to the population variance. These results hold in general and are not a peculiarity of this particular example. We will prove this next.

◀

2.3 Key Results on the Sample Mean, Sample Variance, Sample Minimum, and Sample Maximum

Theorem 2.1 *Let $X_1, X_2, \ldots, X_n$ be a random sample from a distribution with mean μ and variance σ^2. Then $E[\bar{X}] = \mu$ and $Var[\bar{X}] = \dfrac{\sigma^2}{n}$.*

Proof First consider the expectancy

$$E\left[\bar{X}\right] = E\left[\frac{1}{n}\sum_{i=1}^{n} X_i\right] = \frac{1}{n}E\left[\sum_{i=1}^{n} X_i\right]$$

$$= \frac{1}{n}\left[\sum_{i=1}^{n} E\left[X_i\right]\right] = \frac{1}{n}\left[\sum_{i=1}^{n}\mu\right] = \frac{1}{n} \times n\mu = \mu,$$

and now consider the variance

$$Var\left[\bar{X}\right] = Var\left[\frac{1}{n}\sum_{i=1}^{n}X_i\right] = \frac{1}{n^2}Var\left[\sum_{i=1}^{n}X_i\right] = \frac{1}{n^2}\left[\sum_{i=1}^{n}Var\left[X_i\right]\right]$$

$$= \frac{1}{n^2}\left[\sum_{i=1}^{n}\sigma^2\right] = \frac{1}{n^2}\times n\sigma^2 = \frac{\sigma^2}{n},$$

where we can interchange the "Var" operator and summation as we have taken a random sample and so the random variables are independent. $\square$

This result tells us that we expect the sample mean to equal the population mean, and that the amount the sample mean will vary depends on the population variance and the size of the sample taken. We can make the variance of the sample mean smaller by taking a larger sample.

Theorem 2.2 *Let* $X_1, X_2, \ldots, X_n$ *be a random sample from a distribution with mean* μ *and variance* σ^2. *Then* $E[S^2] = \sigma^2$.

Proof We will proceed with the proof using the following expression for the sample variance:

$$S^2 = \frac{1}{n-1}\left\{\sum_{i=1}^{n}X_i^2 - n\bar{X}^2\right\}.$$

It follows that

$$E[S^2] = \frac{1}{n-1}E\left[\left\{\sum_{i=1}^{n}X_i^2 - n\bar{X}^2\right\}\right] = \frac{1}{n-1}\left[\left\{\sum_{i=1}^{n}E\left[X_i^2\right] - nE\left[\bar{X}^2\right]\right\}\right]$$

since the X_i's are independent and identically distributed as they are from a random sample.

Now $Var[X_i] = E[X_i^2] - (E[X_i])^2$ which implies $\sigma^2 = E[X_i^2] - \mu^2$. Similarly $Var[\bar{X}] = E[\bar{X}^2] - (E[\bar{X}])^2$ which implies $\frac{\sigma^2}{n} = E[\bar{X}^2] - \mu^2$ (using Theorem 2.1).

Hence

$$E[S^2] = \frac{1}{n-1}\left\{\sum_{i=1}^{n}(\sigma^2 + \mu^2) - n\left(\frac{\sigma^2}{n} + \mu^2\right)\right\}$$

$$= \frac{1}{n-1}\left\{n\sigma^2 + n\mu^2 - \sigma^2 - n\mu^2\right\} = \frac{(n-1)\sigma^2}{n-1} = \sigma^2.$$ $\square$

Recall that when we defined the sample variance in Definition 2.9, it was mentioned that it may look peculiar that we divide by $(n - 1)$ instead of n. The reason for dividing by $(n - 1)$ can be seen above. It is natural to think it desirable to expect the sample variance to be equal to the population variance, particularly if we intend to use the sample variance to guess the value of the population variance. If we didn't divide by $(n - 1)$ and instead divide by n, we wouldn't have this property (although for large n the difference would be negligible).

R Example

We will consider a simulated example to illustrate Theorems 2.1 and 2.2. Let us suppose we have a population of the number of mobile phones within 15 houses in a particular street in the UK.

We can load this population data into R as follows:

```
pop <- c(3, 2, 1, 1, 1, 2, 2, 2, 1, 3, 4, 1, 2, 3, 2)
```

This is already a rather artificial example, as the population is usually unknown and we take samples from this population in order to try to understand it. The population mean and population variance for this data are $\mu = 2$ and $\sigma^2 = 0.8$, respectively. These are population parameters which would be typically unknown.

In this exercise, we will take 10000 random samples from our population, of sizes $1, 2, \ldots, 15$, and for each of the samples compute its sample mean and sample variance. The following code will allow us to do this exercise:

```
sampmean <- matrix(nrow = 10000, ncol = 15)
sampvar <- matrix(nrow = 10000, ncol = 15)

for (i in 1:10000){
  for (j in 1:15){
    samp <- sample(pop, j, replace = TRUE)
    sampmean[i,j] <- mean(samp)
    sampvar[i,j]  <- var(samp)
  }
}
```

We can obtain boxplots of the results showing the distribution of the sample means and variances for each of the different sample sizes by using

```
boxplot(sampmean, main = 'Distribution of Sample Mean',
          xlab = 'Sample Size',ylab = 'Sample Mean')
boxplot(sampvar, main = 'Distribution of Sample Variance',
          xlab = 'Sample Size', ylab = 'Sample Variance')
```

which produces the following:

Distribution of Sample Mean

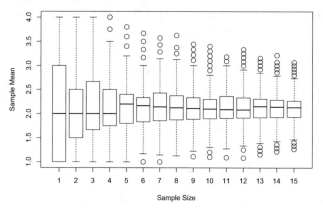

Distribution of Sample Variance

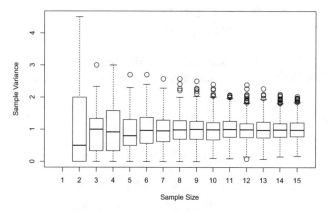

We observe the following:

- As the sample size increases, the amount of variability in both the sample mean and sample variance reduces.
- For the sample mean, the variability is described by Theorem 2.1 and is σ^2/n which is $0.8/n$ for this example. As the sample size gets bigger, the variability in the sample mean gets smaller.
- The sample variance in Definition 2.9 has $n - 1$ on its denominator, which means it is undefined for $n = 1$, and hence no values appear on the figure above for this sample size. This is also intuitive. If there is only one thing in the sample, then how can we talk about variance?

◄

Theorem 2.3 *Let $X_1, X_2, \ldots, X_n$ be a random sample from a continuous distribution with cumulative distribution function $F(x)$ and probability density function $f(x)$. The probability density function of the sample maximum $Z = \max(X_1, X_2, \ldots, X_n)$ is given by*

$$g(z) = nf(z)[F(z)]^{n-1},$$

and the probability density function of the sample minimum $W = \min(X_1, X_2, \ldots, X_n)$ is given by

$$h(w) = nf(w)[1 - F(w)]^{n-1}.$$

Proof Consider $Z = \max(X_1, X_2, \ldots, X_n)$ first. It is easier to proceed with finding the cumulative distribution function of Z, and then differentiate it to obtain its probability density function. Denote the cumulative distribution function of the sample maximum by $G(z)$. Then

$$G(z) = P[Z \le z] = P[X_1 \le z, X_2 \le z, \ldots, X_n \le z]$$
$$= P[X_1 \le z]P[X_2 \le z] \ldots P[X_n \le z] = (F(z))^n,$$

because the X_i's are independent and identically distributed.

The probability density function of the sample maximum is thus given by

$$g(z) = \frac{dG(z)}{dz} = \frac{d\,(F(z))^n}{dz} = nF(z)^{n-1}\frac{dF(z)}{dz} = nf(z)\,(F(z))^{n-1}.$$

Now consider the sample minimum $W = \min(X_1, X_2, \ldots, X_n)$. It follows that

$$P[W \ge w] = P[X_1 \ge w, X_2 \ge w, \ldots, X_n \ge w]$$
$$= P[X_1 \ge w]P[X_2 \ge w] \ldots P[X_n \ge w]$$
$$= (1 - P[X_1 \le w])\,(1 - P[X_2 \le w]) \ldots (1 - P[X_n \le w])$$
$$= (1 - F(w))^n$$

and we can write the cumulative distribution function of the sample minimum as $H(w) = P[W \le w] = 1 - (1 - F(w))^n$.

The probability density function of the sample minimum is thus given by

$$h(w) = \frac{dH(w)}{dw} = \frac{d\left(1 - (1 - F(w))^n\right)}{dw}$$
$$= -n\,(1 - F(w))^{(n-1)}\frac{d\,(1 - F(w))}{dw}$$
$$= nf(w)\,(1 - F(w))^{n-1}. \qquad \square$$

It is possible to prove a similar result for discrete distributions, but we will not do that here. It is interesting to compare and contrast the statement of Theorem 2.3 with Theorems 2.1 and 2.2. Note that in the latter there is no specific need to know the distribution of the population. As long as we have a random sample, we will always expect the sample mean to equal the population mean (and so on). However, we are unable to make such grand statements about the sample maximum or minimum. The properties of these statistics also depend on the distribution of the population—we need to know its probability density function and cumulative distribution function.

Example

The random variable X has probability density function

$$f(x) = 12x^2(1 - x), \quad 0 \le x \le 1.$$

Obtain the probability density function and cumulative distribution function of the sample maximum, when a random sample of size 3 is taken from X. Find the probability that the largest maximum is $\frac{1}{2}$.

Solution: Here the cumulative distribution function of the random variable X is given by

$$F(x) = \int_0^x 12x^2(1 - x)\, dx = x^3(4 - 3x),$$

and so we can find the probability density function of the sample maximum $Z = \max(X_1, X_2, \ldots, X_n)$ to be

$$g(z) = nf(z)\,(F(z))^{n-1} = 36z^2(1 - z)\left(z^3(4 - 3z)\right)^2,$$

where $0 \le z \le 1$. The cumulative distribution function is given by

$$G(z) = P[Z \le z] = \left(z^3(4 - 3z)\right)^3,$$

and the probability that the largest maximum is $\frac{1}{2}$ is

$$G\left(\frac{1}{2}\right) = P\left[Z \le \frac{1}{2}\right] = \left\{\left(\frac{1}{2}\right)^3\left(4 - \frac{3}{2}\right)\right\}^3 = \left(\frac{5}{16}\right)^3 \approx 0.031.$$

◀

2.4   Four Commonly Used Distributions for Populations

In this section, we will describe four commonly used distributions for populations. What follows forms the basis for many of the remaining ideas within this book. Much of classical statistical inference is based on the use and manipulation of the following four distributions.

2.4.1   Normal Distribution

It is assumed that you have some prior experience of this distribution, but some essential details are given here. Let X be a Normally distributed random variable. The Normal distribution is parameterized by the mean and variance of X. We introduce the notation $E[X] = \mu$ and $Var[X] = \sigma^2$. For brevity to encapsulate all of this information, we write $X \sim N[\mu, \sigma^2]$ which can be read as shorthand for "X is a random variable following a Normal distribution with mean μ and variance σ^2." Graphs of this probability density function for different μ and σ^2 are given in Fig. 2.1.

If X is Normally distributed with mean μ and variance σ^2, the so-called standardized random variable $Z = \dfrac{X - \mu}{\sigma}$ is also Normally distributed, with mean 0 and variance 1. We can prove these facts as follows:

$$E[Z] = E\left[\frac{X - \mu}{\sigma}\right] = \frac{E[X] - \mu}{\sigma} = \frac{\mu - \mu}{\sigma} = 0\,,$$

and for the variance

$$Var[Z] = Var\left[\frac{X - \mu}{\sigma}\right] = \frac{Var[X]}{\sigma^2} = \frac{\sigma^2}{\sigma^2} = 1\,.$$

Fig. 2.1 Graphs of the probability density function of the Normal distribution for different μ and σ

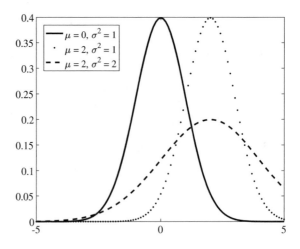

The probabilities and percentiles of a Normal distribution with mean 0 and variance 1 are tabulated (see Table C.1 in the Appendix). We call $Z \sim N[0, 1]$ the *standard Normal distribution*. We will use this as general notation throughout the book: once we have a random variable with a standard Normal distribution, we will denote it by Z.

Example

Let $X \sim N[4, 3^2]$. Find

(i) $P[X < 10]$,
(ii) the value of x such that $P[X < x] = 0.95$.

Solution: We could use a computer to directly obtain the solutions to (i) and (ii). Alternatively, we can consult statistical tables which tabulate probabilities for the standard Normal distribution. If we use statistical tables, we need to convert the quantities (i) and (ii) into corresponding versions for the standard Normal distribution.

We can do this as follows:

$$P[X < 10] = P\left[\frac{X - \mu}{\sigma} < \frac{10 - \mu}{\sigma}\right] = P\left[Z < \frac{10 - 4}{3}\right] = P[Z < 2].$$

We find the value of this final probability using Table C.1. We see that

$$P[Z < 2] = 0.977725.$$

Finally,

$$P[X < x] = 0.95 \implies P\left[\frac{X - \mu}{\sigma} < \frac{x - \mu}{\sigma}\right] = 0.95 \text{ or } P\left[Z < \frac{x - \mu}{\sigma}\right] = 0.95.$$

Values on the standard Normal distribution such that there is an ascribed amount of probability "to the left" of it are tabulated (see Table C.2 in the Appendix), and it can be seen that $P[Z < 1.6449] = 0.95$. We can use this to solve for x, as $\frac{x - 4}{3} = 1.6449$ giving $x = 8.9347$.

◀

At this point, it is useful to introduce some general notation. Let z_γ be the scalar such that $P[Z < z_\gamma] = \gamma$ where Z is a random variable that has a standard Normal distribution, $Z \sim N[0, 1]$. In the above example we have identified that $z_{0.95} = 1.6449$. Other commonly used values include $z_{0.9} = 1.2816$, $z_{0.975} = 1.96$, and $z_{0.99} = 2.3263$. As well as consulting Table C.2 in the Appendix, such values can be found using R with the command qnorm(0.95) which would report the value $z_{0.95}$, for example.

The Normal distribution has the nice property that sums of Normal distributions are also Normally distributed. The theorem stating this is given next.

Theorem 2.4 *If $X_1, X_2, \ldots, X_n$ are independent Normally distributed random variables, such that each X_i has mean μ_i and variance σ_i^2, then for any constants a_i, $i = 1, 2, \ldots, n$, the random variable $\sum_{i=1}^{n} a_i X_i$*

(a) is Normally distributed,

(b) has mean $\sum_{i=1}^{n} a_i \mu_i$,

(c) has variance $\sum_{i=1}^{n} a_i^2 \sigma_i^2$.

This tells us that "a linear combination of Normally distributed variables is also Normally distributed." A specific case of this theorem relating to the sample mean (where $a_i = 1/n$, $\mu_i = \mu$, and $\sigma_i^2 = \sigma^2$ for $i = 1, \ldots, n$) is given below.

Corollary 2.1 *If $X_1, X_2, \ldots, X_n$ are independent Normally distributed random variables, such that each X_i has mean μ and variance σ^2, then*

$$\bar{X} \sim N\left[\mu, \frac{\sigma^2}{n}\right]$$

and we can standardize this random variable as follows:

$$Z = \frac{\bar{X} - \mu}{\sigma/\sqrt{n}} \sim N[0, 1].$$

Note that the above corollary contains some information we already know. Regardless of distribution, for any random sample, we have already proved that $E[\bar{X}] = \mu$ and $Var[\bar{X}] = \frac{\sigma^2}{n}$ (see Theorem 2.1). If we additionally assume that our population is also Normally distributed, we obtain the additional information that the sample mean is also Normally distributed.

Example

The weights of sacks of potatoes are Normally distributed with mean 25 kg and standard deviation 1 kg. Find
(i) the probability that the mean weight of a random sample of four sacks is greater than 26 kg;
(ii) the sample size n necessary for the sample mean to be within 0.25 kg of the true mean 25 kg at least 95% of the time.

Solution: Let X denote the random variable of weights of sacks of potatoes. Then

$$X \sim N[25, 1].$$

(i) Using Corollary 2.1 it follows that $\bar{X} \sim N[25, 1/4]$. Hence

$$P[\bar{X} > 26] = P\left[Z > \frac{26 - 25}{\sqrt{1/4}}\right] = P[Z > 2] = 1 - P[Z \leq 2] = 1 - 0.97725 = 0.02275.$$

(ii) Again using Corollary 2.1, it follows that $\bar{X} \sim N[25, 1/n]$. We wish to find the sample size n such that

$$P[-0.25 < \bar{X} - 25 < 0.25] \geq 0.95.$$

We proceed as follows:

$$P\left[\frac{-0.25}{\sqrt{1/n}} < \frac{\bar{X} - 25}{\sqrt{1/n}} < \frac{0.25}{\sqrt{1/n}}\right] \geq 0.95$$

$$\text{or } P\left[\frac{-0.25}{\sqrt{1/n}} < Z < \frac{0.25}{\sqrt{1/n}}\right] \geq 0.95$$

$$\text{or } P\left[Z < \frac{0.25}{\sqrt{1/n}}\right] - P\left[Z < \frac{-0.25}{\sqrt{1/n}}\right] \geq 0.95$$

$$\text{or } P\left[Z < \frac{0.25}{\sqrt{1/n}}\right] - \left(1 - P\left[Z < \frac{0.25}{\sqrt{1/n}}\right]\right) \geq 0.95$$

$$\text{or } 2P\left[Z < \frac{0.25}{\sqrt{1/n}}\right] \geq 1.95$$

$$\text{or } P\left[Z < \frac{0.25}{\sqrt{1/n}}\right] \geq 0.975$$

From statistical tables $z_{0.975} = 1.96$, that is, $P[Z < 0.975] = 1.96$, where Z follows the standard Normal distribution. Hence $\dfrac{0.25}{\sqrt{1/n}} \geq 1.96$ or $n \geq 61.47$.

◀

2.4.2 Student's t-Distribution

The Student's t-distribution is symmetric and bell-shaped, like the Normal distribution, but has heavier tails, meaning that it is more prone to producing values that fall far from its mean. Let the random variable X have the Student's t-distribution. Its probability density function is given by

$$f(x; v) = \frac{\Gamma(\frac{v+1}{2})}{\sqrt{v\pi}\,\Gamma(\frac{v}{2})}\left(1 + \frac{x^2}{v}\right)^{-\frac{v+1}{2}}.$$

Fig. 2.2 Graphs of the probability density function of the Student's t-distribution for different degrees of freedom ν

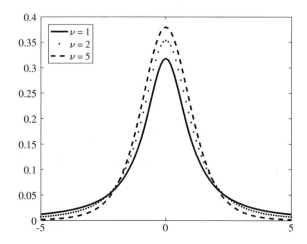

It depends on one parameter ν, called the *degrees of freedom*. For brevity, if X follows the Student's t-distribution with ν degrees of freedom, we write $X \sim t(\nu)$. Note that as $\nu \to \infty$ the Student's t-distribution behaves like the Normal distribution. Plots of the probability density function for the Student's t-distribution for different ν are given in Fig. 2.2.

The formal construction of the Student's t-distribution is given below.

Lemma 2.1 *If $X_1, X_2, \ldots, X_n$ is a random sample from a Normal distribution with mean μ, variance σ^2, and S is the sample standard deviation, then the random variable $T = \dfrac{\bar{X} - \mu}{S/\sqrt{n}}$ has the Student's t-distribution with $(n - 1)$ degrees of freedom. For brevity we write $T \sim t(n - 1)$.*

As before we will use the following notation. Let $t_\gamma(n - 1)$ be the scalar such that $P[T < t_\gamma(n - 1)] = \gamma$, where $T \sim t(n - 1)$. Most statistical tables only give percentiles of T. For example, with $n = 11$, $P[T < 1.812] = 0.95$, and so in the notation introduced $t_{0.95}(10) = 1.812$. Percentage points of the Student's t-distribution are given in Table C.4. Such values can also be found using R with the command qt(0.95, df = 10) which would report the value $t_{0.95}(10)$, for example.

2.4.3 Chi-Squared Distribution

For n a positive integer, the chi-squared distribution is the distribution of the sum

$$X_1^2 + X_2^2 + \ldots + X_n^2$$

where $X_1, X_2, \ldots, X_n$ are all independently and identically distributed standard Normal random variables.

Fig. 2.3 Graphs of the probability density function of the chi-squared distribution for different degrees of freedom ν

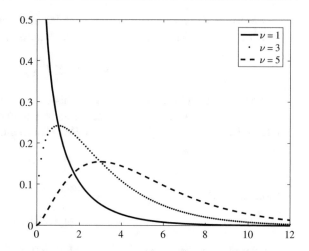

The chi-squared distribution depends on one parameter, which again is called the degrees of freedom. Shortly, if X is a random variable with chi-squared distribution with ν degrees of freedom, we can write $X \sim \chi^2(\nu)$. Here $E[X] = \nu$ and $Var[X] = 2\nu$. Note that this distribution is not symmetric. A plot of the chi-squared distribution for different degrees of freedom ν is given in Fig. 2.3.

The formal construction of the chi-squared distribution is given below.

Lemma 2.2 *Let $X_1, X_2, \ldots, X_n$ be a random sample from a Normal distribution with mean μ, variance σ^2, and sample variance S^2. Then $\chi^2 = \dfrac{(n-1)S^2}{\sigma^2}$ has a chi-squared distribution with $(n-1)$ degrees of freedom. Briefly, we write $\chi^2 \sim \chi^2(n-1)$.*

Again we will use the following notation. Let $\chi^2_\gamma(n-1)$ be the scalar such that $P[\chi^2 < \chi^2_\gamma(n-1)] = \gamma$, where $\chi^2 \sim \chi^2(n-1)$. For example, with $n = 11$ and $\gamma = 0.95$, $\chi^2_{0.95}(10) = 18.307$. This means with a sample size of 11, $P[\chi^2 < 18.307] = 0.95$. Percentage points of the chi-squared distribution are given in Table C.3. Such values can also be found using R with the command `qchisq(0.95, df = 10)` which would report the value $\chi^2_{0.95}(10)$, for example.

Example

Suppose a point (X, Y) is placed at a random location on a side of A4 paper, where X and Y are independent random variables each with the standard Normal distribution. What would the radius of a circle, centered at the origin, drawn on the paper, need to be so that the point (X, Y) has probability 0.99 of laying within it?

Solution: The equation of a circle centered at the origin with radius r is given by $x^2 + y^2 = r^2$. The point (X, Y) will lay within this circle if

$$X^2 + Y^2 \leq r^2.$$

The random variable $X^2 + Y^2$ follows a chi-squared distribution with two degrees of freedom since X and Y are independent, each with the standard Normal distribution. Let $\chi^2 = (X^2 + Y^2) \sim \chi^2(2)$. We wish to find the radius r such that

$$P\left[\chi^2 \leq r^2\right] = 0.99.$$

From Table C.3,

$$\chi^2_{0.99}(2) = 9.210,$$

and this directly means that $P\left[\chi^2 \leq 9.210\right] = 0.99$. The required radius can thus be found as $r = \sqrt{9.210} \approx 3.035$.

◄

2.4.4 F-Distribution

The F-distribution is a skewed distribution which depends on two parameters, v_1 and v_2, both known as degrees of freedom. If $X \sim \chi^2(v_1)$ and $Y \sim \chi^2(v_2)$ are two independent random variables each with the chi-squared distribution, then $F = \dfrac{X/v_1}{Y/v_2}$ is said to have the F-distribution with v_1 and v_2 degrees of freedom. Briefly we write $F \sim F(v_1, v_2)$. Note that if $F \sim F(v_1, v_2)$ then $1/F \sim F(v_2, v_1)$.

Lemma 2.3 *Let $X_1, X_2, \ldots, X_m$ and $Y_1, Y_2, \ldots, Y_n$ be random samples from $X \sim N[\mu_X, \sigma_X^2]$ and $Y \sim N[\mu_Y, \sigma_Y^2]$, respectively. Let S_X^2 and S_Y^2 be the corresponding sample variances. It follows that $\dfrac{(m-1)S_X^2}{\sigma_X^2} \sim \chi^2(m-1)$ and $\dfrac{(n-1)S_Y^2}{\sigma_Y^2} \sim \chi^2(n-1)$. Hence*

$$\frac{S_X^2/\sigma_X^2}{S_Y^2/\sigma_Y^2} \sim F(m-1, n-1).$$

We define $F_\gamma(m-1, n-1)$ to be the scalar such that $P[F < F_\gamma(m-1, n-1)] = \gamma$, where $F \sim F(m-1, n-1)$ and percentage points of the F-distribution are given in Tables C.5, C.6, C.7 and C.8. Such values can also be found using R with the command qf(0.95, df1 = 10, df2 = 5) which would report the value $F_{0.95}(10, 5) = 4.735$, for example. A plot of the F-distribution with different degrees of freedom is given in Fig. 2.4.

Fig. 2.4 Graphs of the probability density function of the F-distribution for different degrees of freedom ν_1 and ν_2

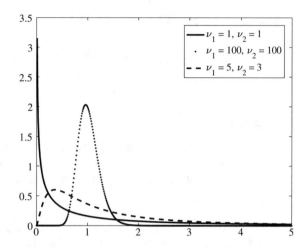

2.5   Central Limit Theorem

Recall Corollary 2.1, which said that if we had a random sample of Normally distributed random variables (which means we have an i.i.d. sample of Normal random variables), then their sample mean is also Normally distributed. The big question is, what if my independent and identically distributed random variables are not Normally distributed? This question gives rise to one of the most important theorems in the whole of statistics, known as the Central Limit Theorem, and is described below.

Theorem 2.5 *Let $X_1, X_2, \ldots, X_n$ be independent and identically distributed random variables with mean μ and variance σ^2. Then $\bar{X} \sim N\left[\mu, \dfrac{\sigma^2}{n}\right]$, approximately.*

Compare the Central Limit Theorem with Corollary 2.1. In the latter, our random variables had to be Normally distributed. This requirement is removed from the Central Limit Theorem. Note though that the result in Corollary 2.1 is exact, whilst we end the statement of the Central Limit Theorem above with the word "approximately". This approximation gets better as n grows. The Central Limit Theorem holds for independent and identically distributed random variables following any distribution.

Example

The duration of a pregnancy is known to have mean 266 days and standard deviation 16 days. In a particular hospital, a random sample of 25 pregnant women was studied. What is the probability that the mean duration of the pregnancy of these 25 women is less than 270 days?

Solution: We can use the Central Limit Theorem to approximate the required probability. Let $\bar{X}$ denote the sample mean of the random sample of 25 women.

Then $\bar{X} \sim N(266, \frac{16^2}{25})$ approximately. We can thus compute

$$P[\bar{X} < 270] = P\left[\frac{\bar{X} - \mu}{\sigma/\sqrt{n}} < \frac{270 - 266}{16/\sqrt{25}}\right] = P[Z < 1.25] = 0.89435.$$

◀

R Example

We will consider an example to illustrate the Central Limit Theorem. Suppose we roll a twice dice 10000 times and plot a histogram of the average score of the two rolls. This can be done using the snippet of code below:

```
RollTwo <- c()
for (i in 1:10000) {
   RollTwo[i] = mean(sample(1:6, 2, replace = TRUE))}
hist(RollTwo, col = "pink", main = "Number of rolls = 2",
                      xlab = "Outcome", prob=TRUE)
curve(dnorm(x, mean(RollTwo), sd(RollTwo)), col="blue",
                      lwd=2, add=TRUE)
```

The histogram looks as follows. We have also overlaid a plot of the probability density function of the Normal distribution.

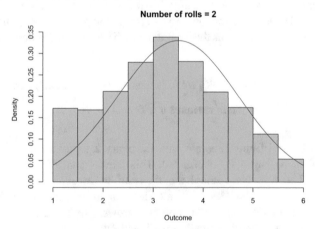

The Central Limit Theorem states that the distribution of the sample mean will tend toward a Normal distribution as the sample size gets larger. In other words, the Normal approximation to the histogram will get better as the number of the rolls of the die increases. Let us visualize this by increasing the number of rolls to 5, 10, and 20. We do each number of rolls 10000 times. This can be done using the code below:

```
RollFive <- c()
RollTen <- c()
RollTwenty <- c()

for (i in 1:10000){
  RollFive[i] = mean(sample(1:6, 5, replace = TRUE))
  RollTen[i] = mean(sample(1:6, 10, replace = TRUE))
  RollTwenty[i] = mean(sample(1:6, 20, replace = TRUE))
}
par(mfrow=c(1,3))
hist(RollFive, col ="green",main="Rolls = 5",
                    prob=TRUE, ylim=c(0,1), xlab="Outcome")
curve(dnorm(x, mean(RollFive), sd(RollFive)), col="blue",
                    lwd=2, add=TRUE)

hist(RollTen, col ="light blue", main="Rolls = 10",
                    prob=TRUE, ylim=c(0,1),  xlab="Outcome")
curve(dnorm(x, mean(RollTen), sd(RollTen)), col="blue",
                    lwd=2, add=TRUE)

hist(RollTwenty, col ="orange",main="Rolls = 20",
                    prob=TRUE, ylim=c(0,1), xlab="Outcome")
curve(dnorm(x, mean(RollTwenty), sd(RollTwenty)), col="blue",
                    lwd=2, add=TRUE)
```

This code produces the figure below.

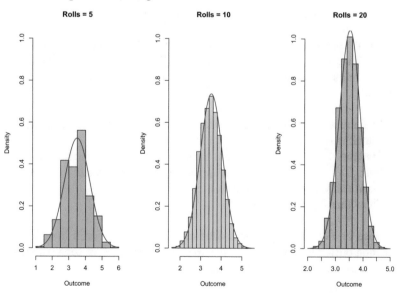

Even for a small number of rolls of the dice, the histogram of the average score looks to follow a Normal distribution.

◀

2.6   Exercises

1. Let X and Y be two independent Normally distributed random variables. X has a mean of 15 and a variance of $\frac{1}{4}$, while Y has a mean of 9 and a variance of $\frac{8}{9}$. Find the probability that $2X - 3Y < 6$, and the probability that $2X - 3Y \geq 6$.

2. Wood screws are manufactured in their millions but packed in boxes of 200 to be sold to the public. If the length of the screws is known to be Normally distributed with a mean of 2 cm and variance of 0.05 cm, calculate the probability that the sample mean length of screws in a box of 200 is greater than 2.02 cm.

3. A small block of flats contains six families. One family has no children, one family has one child, three families have two children, and one family has three children. A random sample of three families is selected.

   a. List all the possible samples of three families according to the number of children.

   b. Calculate the sample mean, sample variance, and sample median for each possible sample.

   c. Find the sampling distributions of the sample mean, variance, and median.

4. Two youth clubs have 10 members, one youth club has 30 members, and two other youth clubs have 40 members. A random sample of two of these youth clubs is taken with replacement. Find the sampling distributions of the sample mean and sample minimum of the number of members. Find the mean of each of these sampling distributions.

5. Before the invention of dice as we know them, the heel bones of sheep were used as gambling tools. They were called astragali and, when tossed, could land on any of four sides. Each side was equally likely and represented scores of 1, 3, 5, and 7.

   a. List all the possible outcomes when two astragali are thrown, and obtain the sampling distribution of the mean score, $\overline{X}$.

   b. Find the sampling distribution of the sample variance of the scores, S^2.

   c. Find $E\left[\overline{X}\right]$, $E\left[S^2\right]$ and $Var\left[\overline{X}\right]$. Explain how these relate to μ and σ^2, the mean and variance of the scores 1, 3, 5, and 7.

   d. Show that $E[S^2] \neq (E[S])^2$.

6. A piece of electrical equipment consists of five bulbs placed in series. It can function only if all the bulbs are working. The lifetime T of each bulb (in units of 1000 hours) has the probability density function

$$f(t) = \frac{2}{(1+t)^3}, \quad 0 \leq t < \infty.$$

Calculate the mean lifetime of the piece of equipment.

7. The time in minutes needed to collect the tolls from motorists crossing a toll bridge has the probability density function

$$f(x) = 2\exp(-2x), \quad 0 \le x < \infty.$$

A motorist approaches the bridge and counts 50 vehicles waiting in a queue to pay the toll. Only one tollbooth is in operation. Use the Central Limit Theorem to find the approximate probability that a motorist will have to wait more than 25 min before reaching the front of the queue. State any assumptions that you make.

Toward Estimation

3.1 Introduction

Suppose we are given a random variable X, which we know is Normally distributed. The Normal distribution is parameterized by its mean μ and its variance σ^2. If we want to compute, for example, $P[X \leq 4]$, then we need to know what μ and σ^2 are. As you may be able to imagine, in many real-life situations these population parameters wouldn't be known. So the question becomes that of *estimating* μ and σ^2.

This chapter concerns estimation of unknown population parameters. Upon taking a sample, we can compute sample statistics and use these to estimate unknown population parameters of interest. Sometimes there may be several alternative estimators that can be used. We therefore need criteria to compare estimators to decide which are the "best". We will suppose that there is some univariate parameter θ that we wish to estimate. Let us denote an estimator of θ by $\hat{\theta}$.

3.2 Bias and Variance

Definition 3.1 An estimator of θ, given by $\hat{\theta}$, is *unbiased* if $E[\hat{\theta}] = \theta$; otherwise, it is *biased* with $bias[\hat{\theta}] = E[\hat{\theta}] - \theta$.

We say an estimator is unbiased if we expect it to equal the very thing that we are trying to estimate. Suppose we have a random sample $X_1, X_2, \ldots, X_n$. In Chap. 2, we derived the following:

- $E[\bar{X}] = \mu$,
- $E[S^2] = \sigma^2$.

© Springer Nature Switzerland AG 2020

J. Gillard, *A First Course in Statistical Inference*,

Springer Undergraduate Mathematics Series,

https://doi.org/10.1007/978-3-030-39561-2_3

Hence, we have already proved that the sample mean is an unbiased estimator of the population mean. Likewise, we expect the sample variance to equal the population variance and thus we have already proved that the sample variance is an unbiased estimator of the population variance. These results hold regardless of the underlying distribution of our data as long as we take random samples from our population.

Example

Let X_1, X_2, X_3, X_4 be a random sample from a distribution having unknown mean μ and variance σ^2. The following estimators for μ have been proposed:

$$\hat{\mu}_1 = (X_1 + X_2 + X_3 + X_4)/4$$
$$\hat{\mu}_2 = X_1 - X_2 + X_3 + X_4$$
$$\hat{\mu}_3 = (2X_1 + 3X_2)/5$$
$$\hat{\mu}_4 = (X_1 + X_4)/2.$$

Determine which of the above estimators are unbiased estimators of μ. Find and compare the variance of each of the above estimators in terms of σ^2.

Solution: Let us consider the first estimator.

$$E[\hat{\mu}_1] = E\left[(X_1 + X_2 + X_3 + X_4)/4\right] = (E[X_1] + E[X_2] + E[X_3] + E[X_4])/4$$
$$= (\mu + \mu + \mu + \mu)/4 = \mu.$$

Hence, $\hat{\mu}_1$ is an unbiased estimator of μ. Its variance is given by

$$Var[\hat{\mu}_1] = Var\left[(X_1 + X_2 + X_3 + X_4)/4\right]$$
$$= (Var[X_1] + Var[X_2] + Var[X_3] + Var[X_4])/16 = (\sigma^2 + \sigma^2 + \sigma^2 + \sigma^2)/16$$
$$= \frac{\sigma^2}{4}.$$

Now for the second estimator.

$$E[\hat{\mu}_2] = E\left[(X_1 - X_2 + X_3 + X_4)/4\right] = (E[X_1] - E[X_2] + E[X_3] + E[X_4])/4$$
$$= (\mu - \mu + \mu + \mu)/4 = \mu/2.$$

Hence, $\hat{\mu}_2$ is a biased estimator of μ. Its variance is given by

$$Var[\hat{\mu}_2] = Var\left[(X_1 - X_2 + X_3 + X_4)/4\right]$$
$$= (Var[X_1] - Var[X_2] + Var[X_3] + Var[X_4])/16$$
$$= (\sigma^2 + \sigma^2 + \sigma^2 + \sigma^2)/16 = \frac{\sigma^2}{4}.$$

Notice that although this estimator is biased, it has the same variance as the one considered previously.

Let us consider the third estimator.

$$E[\hat{\mu}_3] = E\left[(2X_1 + 3X_2)/5\right] = (2E[X_1] + 3E[X_2])/5 = (2\mu + 3\mu)/5 = \mu.$$

Hence, $\hat{\mu}_3$ is an unbiased estimator of μ. Its variance is given by

$$Var[\hat{\mu}_3] = Var\left[(2X_1 + 3X_2)/5\right] = (4Var[X_1] + 9Var[X_2])/25$$
$$= (4\sigma^2 + 9\sigma^2)/25 = \frac{13\sigma^2}{25}.$$

As $\frac{13}{25} > \frac{1}{4}$ this estimator has bigger variance than $\hat{\mu}_1$ and $\hat{\mu}_2$. Finally

$$E[\hat{\mu}_4] = E\left[(X_1 + X_4)/2\right] = (E[X_1] + E[X_4])/2 = (\mu + \mu)/2 = \mu.$$

Hence $\hat{\mu}_4$ is an unbiased estimator of μ. Its variance is given by

$$Var[\hat{\mu}_4] = Var\left[(X_1 + X_2)/2\right] = (Var[X_1] + Var[X_4])/4 = (\sigma^2 + \sigma^2)/4 = \frac{\sigma^2}{2}.$$

As $\frac{13}{25} > \frac{1}{2} > \frac{1}{4}$ this estimator has bigger variance than $\hat{\mu}_1$ and $\hat{\mu}_2$ but smaller variance than $\hat{\mu}_3$.

◀

Example

The number of breakdowns per week for a type of computer is a random variable Y having a Poisson distribution with mean μ. A random sample $Y_1, Y_2, \ldots, Y_n$, of observations on the weekly number of breakdowns is available.

1. Suggest an unbiased estimator for μ.
2. The weekly cost of repairing the breakdowns is $C = 3Y + Y^2$. Show that $E[C] = 4\mu + \mu^2$.
3. Find a function of $Y_1, Y_2, \ldots, Y_n$ that is an unbiased estimator for $E[C]$, the weekly cost of repairing the breakdowns.

Solution:

1. The sample mean is an unbiased estimator of the population mean. Hence, the sample mean, which we will call $\bar{Y}$, is a suitable unbiased estimator for μ, as $E[\bar{Y}] = \mu$.
2. $E[C] = E[3Y + Y^2] = 3E[Y] + E[Y^2]$. As Y has a Poisson distribution with mean μ, then $E[Y] = \mu$. To compute $E[Y^2]$, recall the formula $Var[Y] =$

$E[Y^2] - (E[Y])^2$. As Y has a Poisson distribution with mean μ, it also follows that $Var[Y] = \mu$, and so $E[Y^2] = \mu + \mu^2$. We can write $E[C] = 3E[Y] + E[Y^2] = 3\mu + \mu + \mu^2 = 4\mu + \mu^2$.

3. We wish to find an unbiased estimator for $4\mu + \mu^2$. We know that $\bar{Y}$ is an unbiased estimator for μ. That is, $E[\bar{Y}] = \mu$. A good first guess for an unbiased estimator of $4\mu + \mu^2$ is thus $4\bar{Y} + \bar{Y}^2$.

Let us check to see if this estimator is unbiased. $E[4\bar{Y} + \bar{Y}^2] = 4E[\bar{Y}] + E[\bar{Y}^2]$. Using results regarding the mean and variance of a sample mean, we know that $E[\bar{Y}] = \mu$ and $Var[\bar{Y}] = \mu/n$. Hence (using a similar approach to calculate $E[\bar{Y}^2]$ as in part 2.) $E[4\bar{Y} + \bar{Y}^2] = 4\mu + \mu^2 + \mu/n \neq 4\mu + \mu^2$. Our first guess is thus a biased estimator of $4\mu + \mu^2$. We have a bias of μ/n. However, this bias is easily eliminated.

Consider the estimator $4\bar{Y} + \bar{Y}^2 - \bar{Y}/n$. It follows that $E[4\bar{Y} + \bar{Y}^2 - \bar{Y}/n] = 4\mu + \mu^2 + \mu/n - \mu/n = 4\mu + \mu^2$. Hence, $4\bar{Y} + \bar{Y}^2 - \bar{Y}/n$ is an unbiased estimator of $4\mu + \mu^2$.

In the examples above, we saw that some of the estimators were unbiased, and others were biased. We also found expressions for the variance of the estimators. In an ideal situation, we would like unbiased estimators with small variance. This would mean that we would expect our estimator to be close to the unknown parameter without much variation or deviation. However, suppose we had to choose between two estimators of an unknown parameter. One estimator is unbiased, with very large variance. Another estimator is biased, but with small variance. Which estimator should we choose? There are a number of measures which can be computed which considers a balance between bias and variance. One such is the mean square error, defined in the next section.

3.3   Mean Square Error

Definition 3.2 The mean square error (MSE) of an estimator $\hat{\theta}$ of a parameter θ is defined as

$$MSE[\hat{\theta}] = E[(\hat{\theta} - \theta)^2].$$

It isn't immediately obvious that the mean square error is a balance between the bias and the variance of the estimator $\hat{\theta}$. This is proved in the following theorem.

Theorem 3.1 *The mean square error (MSE) of an estimator $\hat{\theta}$ of a parameter θ can be written as*

$$MSE[\hat{\theta}] = Var[\hat{\theta}] + \left\{bias[\hat{\theta}]\right\}^2.$$

Proof We may write

$$
\begin{aligned}
MSE[\hat{\theta}] &= E[(\hat{\theta} - \theta)^2] \\
&= E[(\hat{\theta} - E[\hat{\theta}] + E[\hat{\theta}] - \theta)^2] \\
&= E[(\hat{\theta} - E[\hat{\theta}])^2 + (E[\hat{\theta}] - \theta)^2 + 2(\hat{\theta} - E[\hat{\theta}])(E[\hat{\theta}] - \theta)] \\
&= E[(\hat{\theta} - E[\hat{\theta}])^2] + E[(E[\hat{\theta}] - \theta)^2] + E[2(\hat{\theta} - E[\hat{\theta}])(E[\hat{\theta}] - \theta)]
\end{aligned}
$$

Let us consider each term in the above expression separately. The first term $E[(\hat{\theta} - E[\hat{\theta}])^2]$ is $Var[\hat{\theta}]$. Recall that the expected squared value of a random variable minus its expected value is, by definition, the variance of that random variable. The second term $E[(E[\hat{\theta}] - \theta)^2]$ is the expected value of the bias of $\hat{\theta}$ which we have denoted as $bias[\hat{\theta}]$. The final term

$$
\begin{aligned}
E[2(\hat{\theta} - E[\hat{\theta}])(E[\hat{\theta}] - \theta)] &= 2E\left[\hat{\theta}E[\hat{\theta}] - \hat{\theta}\theta - E[\hat{\theta}]E[\hat{\theta}] + E[\hat{\theta}]\theta\right] \\
&= 2\left[E[\hat{\theta}]E[\hat{\theta}] - E[\hat{\theta}]\theta - E[\hat{\theta}]E[\hat{\theta}] + E[\hat{\theta}]\theta\right] \\
&= 0.
\end{aligned}
$$

Hence, we have proven that $MSE[\hat{\theta}] = Var[\hat{\theta}] + \left\{bias[\hat{\theta}]\right\}^2$.   □

Example

A scientist wants to estimate the volume v of a cuboid whose unequal edges are of lengths x, y, and z. The scientist can estimate the volume using one of two methods.

Method (i): Obtain a direct measurement V of the volume. V can be regarded as a random variable having mean v and variance σ_V^2.
Method (ii): Obtain measurements X, Y, and Z of the unequal edges, and estimate the volume using $\hat{v} = XYZ$. It may be assumed that X, Y, and Z are independent random variables with respective means x, y, and z and common variance σ^2.
   Show that method (ii) gives an unbiased estimate of the volume, but method (i) yields an estimate with smaller mean square error to method (ii) if

$$
\sigma_V^2 < \sigma^6 + \sigma^4(x^2 + y^2 + z^2) + \sigma^2(x^2y^2 + x^2z^2 + y^2z^2).
$$

Solution: Method (i) is unbiased, since $E[V] = v$. For method (ii), as X, Y, and Z are independent then $E[\hat{v}] = E[XYZ] = E[X]E[Y]E[Z] = xyz = v$ and so this method is also unbiased. The mean square error of both these estimators is thus given by their variances. We have already been given the variance of method (i) so it remains to find the variance of method (ii).

$$Var[\hat{v}] = E[\hat{v}^2] - \left\{E[\hat{v}]\right\}^2 = E[X^2Y^2Z^2] - v^2 = E[X^2]E[Y^2]E[Z^2] - v^2 .$$

Now

$$E[X^2] = Var[X] + \{E[X]\}^2 = \sigma^2 + x^2$$
$$E[Y^2] = Var[Y] + \{E[Y]\}^2 = \sigma^2 + y^2$$
$$E[Z^2] = Var[Z] + \{E[Z]\}^2 = \sigma^2 + z^2$$

and so $Var[\hat{v}] = (\sigma^2 + x^2)(\sigma^2 + y^2)(\sigma^2 + z^2) - v^2$. Method (i) will have smaller mean square error than method (ii) if

$$\sigma_V^2 < Var[\hat{v}] = (\sigma^2 + x^2)(\sigma^2 + y^2)(\sigma^2 + z^2) - v^2 ,$$

and expansion of the brackets gives the desired inequality.

◄

Example

The random variable X is Binomially distributed with parameters n (known) and p (unknown). In order to estimate p, the following estimators are proposed:

$$\hat{p}_1 = \frac{X}{n} ,$$

$$\hat{p}_2 = \frac{X + 1}{n + 2} .$$

1. Show that, given n, $\hat{p}_2$ has smaller mean square error than $\hat{p}_1$ provided that p lies in the interval

$$\left(\frac{1}{2} - \frac{\sqrt{(n + 1)(2n + 1)}}{2(2n + 1)} , \frac{1}{2} + \frac{\sqrt{(n + 1)(2n + 1)}}{2(2n + 1)} \right) .$$

2. Evaluate this interval for $n = 1, 2, 3, 4$ and show for large n it approximates to $(0.146, 0.854)$.

Solution:

1. Since $X \sim Bin(n, p)$ then $E[X] = np$ and $Var[X] = npq$ where $q = 1 - p$. Hence

$$E[\hat{p}_1] = \frac{E[X]}{n} = \frac{np}{n} = p .$$

This estimator is unbiased. Its variance is given by

$$Var[\hat{p}_1] = \frac{Var[X]}{n^2} = \frac{npq}{n^2} = \frac{pq}{n} \, .$$

As this estimator is unbiased, its mean square error is also given by its variance and so

$$MSE[\hat{p}_1] = \frac{pq}{n} \, .$$

Let us now consider the second estimator.

$$E[\hat{p}_2] = \frac{E[X]+1}{n+2} = \frac{np+1}{n+2} \, ,$$

and so this estimator is biased with

$$bias[\hat{p}_2] = \frac{np+1}{n+2} - p = \frac{1-2p}{n+2} \, .$$

Its variance is given by

$$Var[\hat{p}_2] = \frac{Var[X]}{(n+2)^2} = \frac{npq}{(n+2)^2}$$

and its mean square error is

$$MSE[\hat{p}_2] = Var[\hat{p}_2] + \{bias[\hat{p}_2]\}^2 = \frac{npq}{(n+2)^2} + \frac{(1-2p)^2}{(n+2)^2} \, .$$

Thus $MSE[\hat{p}_2] < MSE[\hat{p}_1]$ if

$$\frac{npq + (1-2p)^2}{(n+2)^2} < \frac{pq}{n}$$

or if $p^2(8n+4) - p(8n+4) + n < 0$ which is satisfied if p lies in between the two roots of the quadratic equation $p^2(8n+4) - p(8n+4) + n = 0$. Using the quadratic formula, the roots are

$$\frac{1}{2} \pm \frac{\sqrt{(n+1)(2n+1)}}{2(2n+1)}$$

and $MSE[\hat{p}_2] < MSE[\hat{p}_1]$ if p lies in the interval

$$\left(\frac{1}{2} - \frac{\sqrt{(n+1)(2n+1)}}{2(2n+1)}, \frac{1}{2} + \frac{\sqrt{(n+1)(2n+1)}}{2(2n+1)} \right) \, .$$

2. For $n = 1, 2, 3, 4$ the interval end points are shown below:

$$n = 1, \quad (0.09175, 0.90825)$$
$$n = 2, \quad (0.11270, 0.88730)$$
$$n = 3, \quad (0.12204, 0.87796)$$
$$n = 4, \quad (0.12732, 0.87268)$$

Now

$$\frac{\sqrt{(n+1)(2n+1)}}{2(2n+1)} = \frac{1}{2}\sqrt{\frac{n+1}{2n+1}} = \frac{1}{2}\sqrt{\frac{1+1/n}{2+1/n}}$$

which tends to $1/2\sqrt{2}$ as $n \to \infty$. Hence, the interval tends to $1/2 \pm 1/2\sqrt{2} = (0.146, 0.854)$, to three decimal places.

◀

3.4   Exercises

1. The following sample of size 5 was taken randomly from a population with unknown mean and unknown variance: 4, 2, 3, 1, 5. Compute unbiased estimates for the population mean and population variance.
2. A coin, when tossed, has probability p of falling heads. In order to estimate the probability θ of obtaining two heads in two successive tosses, the coin is tossed n times $(n \geq 2)$. If X heads are obtained, show that $\hat{\theta} = \dfrac{X(X-1)}{n(n-1)}$ is an unbiased estimator for θ.
3. A thousand boxes are stored in a warehouse. A quarter of the boxes weigh 2 kg each, half of the boxes weigh 3 kg each, and the remainder weigh 4 kg each. Three boxes are chosen at random (with replacement). Let $\bar{X}$ and M denote the mean and median, respectively, of the weights of these three boxes. Find the sampling distributions of $\bar{X}$ and M, and show that $1000\bar{X}$ and $1000M$ are unbiased estimators of the total weight of all the boxes. Which estimate of the total weight of all the boxes has the smallest mean square error?
4. A firm wishing to estimate the mean number of stoppages per day of a certain piece of machinery decides to make use of past records. These records are not complete and show only for each day whether there were either (i) no stoppages or (ii) at least one stoppage. Examination of these records shows that during a period covering 1000 working days, there were 96 days with no stoppages. Assuming the number of stoppages each day is a Poisson random variable with mean λ, use this information to estimate λ.
5. The random variable X has the Exponential distribution with probability density function

$$f(x; \theta) = \frac{1}{\theta} \exp\left(-\frac{x}{\theta}\right), \quad x > 0, \ \theta > 0.$$

A random sample $X_1, X_2, \ldots, X_n$ is taken from X.

a. Show that the probability density function of the minimum W of the random sample is given by $h(w; \theta) = \dfrac{n}{\theta} \exp\left(-\dfrac{nw}{\theta}\right)$, $w > 0$, $\theta > 0$.

b. Show that W is a biased estimator of θ, and hence propose an alternative estimator, based on W, which is an unbiased estimator of θ.

6. The lifetime T of a bulb has the probability density function

$$f(t) = \alpha^2 t \exp(-\alpha t)$$

for $t \geq 0$, where α is an unknown positive constant. Five bulbs were tested to destruction, and the observed lifetimes, in hours, were

$$870.2, 502.9, 540.3, 62.5, 589.8.$$

Show that the expected value of T is twice the mode (the mode is the value of t which maximizes $f(t)$) and hence find an unbiased estimator of the mode. Find the variance of your estimate.

7. $\hat{\theta}_1$ and $\hat{\theta}_2$ are independent unbiased estimators for an unknown parameter θ, with respective variances σ_1^2 and σ_2^2. Show that $\hat{\theta}_3 = \lambda\hat{\theta}_1 + (1 - \lambda)\hat{\theta}_2$ is also an unbiased estimator of θ for all values of λ. Calculate the variance of $\hat{\theta}_3$ and show that the value of λ that minimizes it is

$$\lambda = \frac{\sigma_2^2}{\sigma_1^2 + \sigma_2^2}.$$

a. Let $\bar{X}_1$ and $\bar{X}_2$ be the means of two independent random samples of sizes n_1 and n_2 from the independent distributions $N[\mu, \sigma_1^2]$ and $N[\mu, \sigma_2^2]$, respectively, where the values of σ_1^2 and σ_2^2 are known. Find the value of λ which minimizes the variance of the estimator $\hat{\mu} = \lambda\bar{X}_1 + (1 - \lambda)\bar{X}_2$.

b. Let S_1^2 and S_2^2 be the sample variances of two independent random samples of sizes n_1 and n_2 from the distributions $N[\mu_1, \sigma^2]$ and $N[\mu_2, \sigma^2]$, respectively. Find the value of λ which minimizes the variance of the estimator $\hat{\sigma}^2 = \lambda S_1^2 + (1 - \lambda)S_2^2$. You may assume that if S^2 is the sample variance of a random sample of size n from the distribution $N[\mu, \sigma^2]$, then $Var[S^2] = \frac{2\sigma^4}{n-1}$.

8. Let X be a random variable with Uniform distribution on $[0, \theta]$, with probability density function $1/\theta$ for $0 \leq x \leq \theta$. Let $X_1, X_2, \ldots, X_n$ be a random sample from X. Show that $Z = \max(X_1, X_2, \ldots, X_n)$ is a biased estimator for θ, and suggest a correction to Z such that the new estimator is unbiased.

Confidence Intervals

4

4.1 Introduction

In Chap. 3, we considered estimating an unknown population parameter denoted θ. The estimator $\hat{\theta}$ is known as a point estimator; it is a single guess as to the value of θ which we know is subject to possible bias and variation. We also discussed that an estimator is a random variable. For example, an unbiased estimator of the population mean is the sample mean. If I take a random sample of observations to compute the sample mean, then it is quite possible upon taking a different random sample that the value of the sample mean (and thus my estimate of the population mean) will change.

Instead of quoting a point estimate, let us aim instead to give a range of possible values in which we are confident includes the true value of the unknown population parameter. Such a range of values is known as a confidence interval. The formal definition follows.

Definition 4.1 A $100(1 - \alpha)\%$ confidence interval for an unknown parameter θ with estimator $\hat{\theta}$ is an interval

$$C = [\hat{\theta} - l, \hat{\theta} + u]$$

with a lower value l and upper value u such that

$$P[\theta \in C] = 1 - \alpha .$$

This interval C is random, because the value of $\hat{\theta}$ depends on the data that has been randomly sampled. In the long run, $100(1 - \alpha)\%$ of confidence intervals will contain θ. An example demonstrating this property is given later.

Confidence intervals are controversial, with many arguments as to their use and benefits. We will visit some of these arguments. Let us begin by deriving a confidence

© Springer Nature Switzerland AG 2020
J. Gillard, *A First Course in Statistical Inference*,
Springer Undergraduate Mathematics Series,
https://doi.org/10.1007/978-3-030-39561-2_4

interval so we can illustrate the concept. Suppose a random variable X is Normally distributed with unknown mean μ and known variance σ^2. If we take a random sample $X_1, X_2, \ldots, X_n$ from X, then we know from Corollary 2.1 that $\bar{X} \sim N[\mu, \sigma^2/n]$, or equivalently, $Z = \dfrac{\bar{X} - \mu}{\sigma/\sqrt{n}} \sim N[0, 1]$.

Using this statement, we are able to write by consulting statistical tables (Table C.2 in this case) that

$$P[-1.96 < Z < 1.96] = P\left[-1.96 < \frac{\bar{X} - \mu}{\sigma/\sqrt{n}} < 1.96\right] = 0.95 .$$

Don't worry at this stage if this statement appears out of the blue. The next section gives more details as to its derivation. At the moment recognize that we have found numerical values $(-1.96, 1.96)$ such that the probability our random variable Z (which depends on the unknown parameter μ) lies between them is 0.95. If we re-arrange the inequality within this probability statement, we can obtain a so-called 95% confidence interval for μ:

$$P\left[\bar{X} - 1.96\frac{\sigma}{\sqrt{n}} < \mu < \bar{X} + 1.96\frac{\sigma}{\sqrt{n}}\right] = 0.95 .$$

Let us make some remarks:

- In our confidence interval above $\bar{X}$ appears. This is a random variable, and so

$$\left[\bar{X} - 1.96\frac{\sigma}{\sqrt{n}}, \bar{X} + 1.96\frac{\sigma}{\sqrt{n}}\right]$$

  is formally a *random interval*. Upon observing a random sample of n observations from X, namely, $x_1, x_2, \ldots, x_n$, we can compute $\bar{x}$ and find numerical values for the end points of this interval by replacing $\bar{X}$ with an observed sample mean. Remember that in this example we assume that the variance σ^2 is known. We say we are 95% confident that this interval contains μ.
- The value 1.96 in the above probability statement resulted in a 95% confidence interval. We can change this value to obtain different levels of confidence. These values can be found from statistical tables. Here is a table of values instead of 1.96 that could be used and the resulting confidence intervals:

Confidence level	Confidence interval
50%	$\bar{x} \pm 0.6745\,\sigma/\sqrt{n}$
68.3%	$\bar{x} \pm \sigma/\sqrt{n}$
90%	$\bar{x} \pm 1.6449\,\sigma/\sqrt{n}$
95%	$\bar{x} \pm 1.96\,\sigma/\sqrt{n}$
99%	$\bar{x} \pm 2.5758\,\sigma/\sqrt{n}$

Here $\bar{x} \pm 0.6745\sigma/\sqrt{n}$ is an abbreviation for $\left[\bar{x} - 0.6745\dfrac{\sigma}{\sqrt{n}}, \bar{x} + 0.6745\dfrac{\sigma}{\sqrt{n}} \right]$.
Studying the above table reveals the following. As my confidence increases (assuming $\bar{x}$, σ and n are fixed), the width of my interval also increases. This makes sense pragmatically. If I was tasked to hit a target with a football, I would be more confident of succeeding with this task if my target were larger rather than smaller. Note also that it is conventional to denote confidence as a percentage, so we often talk of 95% confidence intervals rather than 0.95 confidence intervals. Assuming all else is fixed, for a given confidence level the width of a confidence interval decreases as n increases. This agrees with intuition: our estimate will be more precise if we take a larger sample. The presence of the square root means to half the width of our interval we need to quadruple the sample size.

- You may have noticed that there are some problems with confidence intervals in that they are a different probability statement than those typically considered. For this example μ is a fixed (albeit unknown) parameter and $\bar{X}$ is a random variable. Strictly speaking, we should call $\left[\bar{X} - 1.96\dfrac{\sigma}{\sqrt{n}}, \bar{X} + 1.96\dfrac{\sigma}{\sqrt{n}} \right]$ a "95% confidence interval estimator" for μ and the resulting numerical interval $\left[\bar{x} - 1.96\dfrac{\sigma}{\sqrt{n}}, \bar{x} + 1.96\dfrac{\sigma}{\sqrt{n}} \right]$ a particular "95% confidence interval estimate," but this distinction is rarely highlighted and the term 95% confidence interval refers to both the random interval and its numerical estimate.

This list of subtleties and nuances of confidence intervals is not complete. However, they remain one of the fundamental ideas of statistical inference.

4.2 Commonly Used Confidence Intervals

In the example above, we derived a particular confidence interval for an unknown parameter. The main steps were to construct a probability statement about a random variable depending on that parameter, and then by rearranging this probability statement we could form a confidence interval for that unknown parameter.

The key idea and significant first step in deriving any confidence interval is the writing down of an initial probability statement which can be "solved" for our unknown parameter. Above we considered a 95% confidence interval but generally we can find a $100(1 - \alpha)\%$ confidence interval, where α is given. In the following subsection, we first revisit the above example.

4.2.1   Confidence Intervals for Unknown Means

4.2.1.1 Confidence Interval for an Unknown Mean with Normally Distributed Population and Known Variance

Here, we construct a $100(1 - \alpha)\%$ confidence interval for the unknown mean μ of a Normally distributed population, with known variance σ^2, having observed a random sample $X_1, X_2, \ldots, X_n$ from this population.

As before $Z = \dfrac{\bar{X} - \mu}{\sigma/\sqrt{n}} \sim N[0, 1]$. Our starting point is to find a lower bound l and upper bound u such that

$$P[l < Z < u] = P\left[l < \frac{\bar{X} - \mu}{\sigma/\sqrt{n}} < u\right] = (1 - \alpha)$$

for a given α.

Have a look at the picture below of the (standard Normal) distribution of the random variable Z.

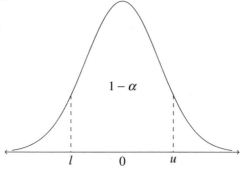

Note that it is symmetric about 0. By inspecting the picture above, one can visualize lots of possible values (there are infinitely many in fact) of l and u so that $P[l < Z < u] = (1 - \alpha)$. We can move l and u so the area under the curve between the dashed lines is $(1 - \alpha)$.

Recall that we defined z_γ to be the value such that $P[Z < z_\gamma] = \gamma$ where $Z \sim N[0, 1]$. For the time being, we will proceed with $l = z_{\frac{\alpha}{2}}$ and $u = z_{1-\frac{\alpha}{2}}$. Then

$$P\left[z_{\frac{\alpha}{2}} < Z < z_{1-\frac{\alpha}{2}}\right] = P\left[Z < z_{1-\frac{\alpha}{2}}\right] - P\left[Z < z_{\frac{\alpha}{2}}\right] = 1 - \frac{\alpha}{2} - \frac{\alpha}{2} = 1 - \alpha .$$

By symmetry we can get the simplification that $l = z_{\frac{\alpha}{2}} = -z_{1-\frac{\alpha}{2}} = -u$. An immediate question is why we have made this choice of l and u. Pragmatically, it simplifies things because our confidence interval will only contain one percentage point from the standard Normal distribution, namely, $z_{1-\frac{\alpha}{2}}$. More importantly, this "symmetric" choice of l and u yields a confidence interval with minimal width compared to any other choice of l and u.

So

$$P\left[-z_{1-\frac{\alpha}{2}} < Z < z_{1-\frac{\alpha}{2}}\right] = P\left[-z_{1-\frac{\alpha}{2}} < \frac{\bar{X} - \mu}{\sigma/\sqrt{n}} < z_{1-\frac{\alpha}{2}}\right] = (1 - \alpha)$$

and by rearranging the inequality we obtain

$$P\left[\bar{X} - z_{1-\frac{\alpha}{2}}\frac{\sigma}{\sqrt{n}} < \mu < \bar{X} + z_{1-\frac{\alpha}{2}}\frac{\sigma}{\sqrt{n}}\right] = (1 - \alpha)$$

giving the $100(1 - \alpha)\%$ confidence interval for μ to be $\bar{x} \pm z_{1-\frac{\alpha}{2}}\dfrac{\sigma}{\sqrt{n}}$ upon replacing the random variable $\bar{X}$ with the sample statistic $\bar{x}$. For a given α, the value $z_{1-\frac{\alpha}{2}}$ can be found from Table C.2.

Example

A random sample of size $n = 6$ gave values $18, 14, 9, 7, 12, 12$ when selected from a Normally distributed population with unknown mean μ and known population variance $\sigma^2 = 24$. Obtain a 95% confidence interval for μ.

Solution: The $100(1 - \alpha)\%$ confidence interval for μ is $\bar{x} \pm z_{1-\frac{\alpha}{2}}\dfrac{\sigma}{\sqrt{n}}$. Here $\alpha = 0.05$ which can be obtained by solving $100(1 - \alpha) = 95$ and from statistical tables $z_{0.975} = 1.96$. The sample mean is computed as $\bar{x} = 12$. The 95% confidence interval is thus given by $12 \pm 1.96\sqrt{\dfrac{24}{6}} = [8.08, 15.92]$.

◀

4.2.1.2 Confidence Interval for the Difference of Two Unknown Means with Normally Distributed Populations and Known Variances

Let us state the confidence interval before discussing how we may derive it.

Let $X_1, X_2, \ldots, X_m$, and $Y_1, Y_2, \ldots, Y_n$, be two random samples, each from independent Normal distributions with unknown means μ_X and μ_Y, but known variances σ_X^2 and σ_Y^2, respectively. A $100(1 - \alpha)\%$ confidence interval for the difference of the unknown means $\mu_X - \mu_Y$ is given by

$$(\bar{x} - \bar{y}) \pm z_{1-\alpha/2}\sqrt{\frac{\sigma_X^2}{m} + \frac{\sigma_Y^2}{n}}.$$

This confidence interval is an extension of that derived previously. Here are the main ideas, which you should verify.

We can begin by writing down that $\bar{X} \sim N\left[\mu_X, \sigma_X^2/m\right]$ and $\bar{Y} \sim N\left[\mu_Y, \sigma_Y^2/n\right]$. Recall from Theorem 2.4 that a linear combination of Normally distributed random variables is also Normally distributed.

Hence $\bar{X} - \bar{Y}$ is Normally distributed with mean

$$E\left[\bar{X} - \bar{Y}\right] = E\left[\bar{X}\right] - E\left[\bar{Y}\right] = \mu_X - \mu_Y,$$

and variance

$$Var\left[\bar{X} - \bar{Y}\right] = Var\left[\bar{X}\right] + Var\left[\bar{Y}\right] = \frac{\sigma_X^2}{m} + \frac{\sigma_Y^2}{n}.$$

In summary

$$(\bar{X} - \bar{Y}) \sim N\left[\mu_X - \mu_Y, \frac{\sigma_X^2}{m} + \frac{\sigma_Y^2}{n}\right]$$

and so upon standardizing

$$Z = \frac{(\bar{X} - \bar{Y}) - (\mu_X - \mu_Y)}{\sqrt{\sigma_X^2/m + \sigma_Y^2/n}} \sim N[0, 1].$$

We can therefore find a value $z_{1-\alpha/2}$ such that $P[-z_{1-\alpha/2} < Z < z_{1-\alpha/2}] = 1 - \alpha$. This equation can be rearranged to yield a $100(1 - \alpha)\%$ confidence interval for $\mu_X - \mu_Y$ which was given earlier. You should verify this formula yourself.

Example

The mean lives of two types of light bulbs, X and Y, were compared by testing 20 random bulbs of type X and 25 random bulbs of type Y. The sample means, in obvious notation, were given by $\bar{x} = 1021.3$ h and $\bar{y} = 1005.7$ h. Assuming that the life of both types of bulbs is Normally distributed with standard deviation 30 h, find a 95% confidence interval for the difference in the population means.

Solution: For a 95% confidence interval $z_{1-\alpha/2} = z_{0.975} = 1.96$. The confidence interval for the difference in population means is

$$(\bar{x} - \bar{y}) \pm z_{1-\alpha/2}\sqrt{\frac{\sigma_X^2}{m} + \frac{\sigma_Y^2}{n}} = (1021.3 - 1005.7) \pm 1.96\sqrt{\frac{900}{20} + \frac{900}{25}} = [-2.04, 33.24].$$

◄

4.2.1.3 Confidence Interval for an Unknown Mean with Normally Distributed Population and Unknown Variance

The previous confidence intervals have assumed that we know the population variance(s). This is quite an unlikely assumption to happen in reality. So, what can we do if we don't know the population variance?

An unbiased estimator of the population variance is the sample variance. To find a confidence interval for an unknown population mean when we use the sample variance to estimate the population variance, we can make use of the following result stated in an earlier chapter (Lemma 2.1).

If $X_1, X_2, \ldots, X_n$ is a random sample from a Normal distribution with mean μ, variance σ^2, and S is the sample standard deviation, then the random variable $T = \dfrac{\bar{X} - \mu}{S/\sqrt{n}}$ has the Student's t-distribution with $(n-1)$ degrees of freedom. For brevity we write $T \sim t(n-1)$.

We can construct a $100(1 - \alpha)\%$ confidence interval for the unknown mean μ of a Normally distributed population with unknown variance, having observed a random sample $X_1, X_2, \ldots, X_n$ analogously to before, because the probability that the random variable T is between $-t_{1-\frac{\alpha}{2}}(n-1)$ and $t_{1-\frac{\alpha}{2}}(n-1)$ is $1 - \alpha$. The derivation is similar to the previous confidence intervals as like the Normal distribution, the Student's t-distribution is symmetrical about zero, yielding the idea and required manipulations identical. We have that

$$P\left[-t_{1-\frac{\alpha}{2}}(n-1) < T < t_{1-\frac{\alpha}{2}}(n-1)\right] = P\left[-t_{1-\frac{\alpha}{2}}(n-1) < \frac{\bar{X} - \mu}{S/\sqrt{n}} < t_{1-\frac{\alpha}{2}}(n-1)\right]$$
$$= (1 - \alpha)$$

and by rearranging the inequality we obtain

$$P\left[\bar{X} - t_{1-\frac{\alpha}{2}}(n-1)\frac{S}{\sqrt{n}} < \mu < \bar{X} + t_{1-\frac{\alpha}{2}}(n-1)\frac{S}{\sqrt{n}}\right] = (1 - \alpha)$$

giving the $100(1 - \alpha)\%$ confidence interval for μ to be $\bar{x} \pm t_{1-\frac{\alpha}{2}}(n-1)\dfrac{s}{\sqrt{n}}$.

4.2.1.4 Confidence Intervals Using the Central Limit Theorem

All of our previously derived confidence intervals for a population mean begin with observing a random sample $X_1, X_2, \ldots, X_n$ from a Normal distribution, and we have separate cases for whether we know the population variance or not. The obvious question to ask is, what if my random sample is not from a Normal distribution?

In this case, we can use the Central Limit Theorem given in Sect. 2.5 to obtain an approximate confidence interval for the population mean. Recall that the Central Limit Theorem is stated as follows:

Let $X_1, X_2, \ldots, X_n$ be independent and identically distributed random variables with mean μ and variance σ^2. Then $\bar{X} \sim N\left[\mu, \dfrac{\sigma^2}{n}\right]$, approximately.

This means that we can use the confidence intervals we have derived previously, to obtain approximate confidence intervals in the situation we cannot assume that our population follows a Normal distribution. We demonstrate the use of the Central Limit Theorem to obtain an approximate confidence interval with an example.

Example

Suppose that the random variable X has a Poisson distribution with unknown mean λ. A random sample of size 100 is taken from the distribution of X. Given that the sample mean of the observed sample was found to be 6.1, obtain an approximate 90% confidence interval for λ.

Solution: The variance of a random variable with Poisson distribution is the same as its mean. We can use the sample mean as an estimate of the population mean and population variance. By the Central Limit Theorem, $\bar{X} \sim N\left[6.1, \dfrac{6.1}{100}\right]$ approximately.

As the variance is estimated, we will use the Student's t-distribution. For a 90% confidence interval, we use $t_{0.95}(99) \approx 1.290$. The 90% confidence interval for λ is

$$\bar{x} \pm t_{1-\frac{\alpha}{2}}(n-1)\sqrt{\frac{s^2}{n}} = 6.1 \pm 1.290\sqrt{\frac{6.1}{100}} = [5.7814, 6.4186].$$

◀

4.2.1.5 Confidence Interval for the Difference of Two Unknown Means with Normally Distributed Populations and Unknown Variances

We now wish to find confidence intervals for the difference of two population means where the population variances are also unknown. Let $X_1, X_2, \ldots, X_m$ and $Y_1, Y_2, \ldots, Y_n$ be two random samples, each from independent Normal distributions with unknown means μ_X and μ_Y and unknown variances σ_X^2 and σ_Y^2, respectively.

We can consider two cases, and below we will take each one in turn:

1. Both population variances are equal, $\sigma_X^2 = \sigma_Y^2 = \sigma^2$. In this case, we assume both samples come from populations which may have equal variances. This means we can use both samples combined to estimate σ^2.
2. Both population variances are unequal, $\sigma_X^2 \neq \sigma_Y^2$. In this case, we assume both samples come from populations which do not have equal variances, and so we have to estimate σ_X^2 and σ_Y^2 separately.

Case 1 : $\sigma_X^2 = \sigma_Y^2 = \sigma^2$

Let S_X^2 be the sample variance of $X_1, X_2, \ldots, X_m$, and let S_Y^2 be the sample variance of $Y_1, Y_2, \ldots, Y_n$. We know that S_X^2 is an unbiased estimator for σ_X^2 and that S_Y^2 is an unbiased estimator for σ_Y^2. As we assume that $\sigma_X^2 = \sigma_Y^2 = \sigma^2$, then both S_X^2 and S_Y^2 are unbiased estimators of σ^2.

The reason why this case is interesting is mainly in the following idea. We have two unbiased estimators of σ^2. Could we combine these estimators in some way to obtain another unbiased estimator, which is better than any one of our original two unbiased estimators? The answer to this question is yes. But, what do we mean by better? In this situation, we mean getting another unbiased estimator which has smaller variance than any of our original two unbiased estimators.

The optimal linear combination of S_X^2 and S_Y^2 to estimate σ^2 (in the sense that it is an unbiased estimator with variance smaller than $Var[S_X^2]$ or $Var[S_Y^2]$) is

$$S^2 = \frac{(m-1)S_X^2 + (n-1)S_Y^2}{m+n-2}.$$

This result was proven in Question 7 from the exercises in Chap. 3. One can show that the random variable

$$T = \frac{(\bar{X} - \bar{Y}) - (\mu_X - \mu_Y)}{S\sqrt{\frac{1}{m} + \frac{1}{n}}}$$

has the Student's t-distribution with $m+n-2$ degrees of freedom. Hence, one may construct a $100(1-\alpha)\%$ confidence interval for $\mu_X - \mu_Y$ as

$$(\bar{x} - \bar{y}) \pm t_{1-\alpha/2}(m+n-2)s\sqrt{\frac{1}{m} + \frac{1}{n}}.$$

Example

The ratio of oxygen consumption of two groups of men was measured. One group (X) trained regularly over a period of time, and the other (Y) trained intermittently. Statistics calculated from the data recorded are given below: Assuming

Group X:	$m = 9$	$\bar{x} = 43.71$	$s_X^2 = 34.5744$
Group Y:	$n = 7$	$\bar{y} = 39.63$	$s_Y^2 = 58.9824$

independent Normal distributions for the readings, and that the variances of the populations are equal, find a 95% confidence interval for the difference between the population means.

Solution: As it is assumed that the variances of the population are equal, then we can pool s_X^2 and s_Y^2 to obtain an unbiased estimator with smaller variance than if we used either of them separately. The optimal combination of s_X^2 and s_Y^2 to estimate the common unknown population variance is

$$s^2 = \frac{(m-1)s_X^2 + (n-1)s_Y^2}{m+n-2} = \frac{(8 \times 34.5744) + (6 \times 58.9824)}{9+7-2} = 45.03497143 \, .$$

One may calculate a 95% confidence interval for the difference in population means as

$$(43.71 - 39.63) \pm t_{0.975}(14)\sqrt{45.03497143}\sqrt{\frac{1}{9}+\frac{1}{7}} = [-3.1742, \, 11.3342] \, ,$$

where $t_{0.975}(14) = 2.145$.

◄

Case 2 : $\sigma_X^2 \neq \sigma_Y^2$

Let S_X^2 be the sample variance of $X_1, X_2, \ldots, X_m$, and let S_Y^2 be the sample variance of $Y_1, Y_2, \ldots, Y_n$. If $\sigma_1^2 \neq \sigma_2^2$ then

$$T = \frac{(\bar{X} - \bar{Y}) - (\mu_X - \mu_Y)}{\sqrt{\frac{S_X^2}{m} + \frac{S_Y^2}{n}}} \tag{4.1}$$

can be shown to have a Student's t-distribution with some degrees of freedom, which we will denote ν. Currently, there is no exact formula for the degrees of freedom ν in the case that both population variances are unknown and cannot be assumed equal.

The so-called Welch–Satterthwaite equation is used to calculate an approximation to the effective degrees of freedom of a linear combination of independent sample variances, also known as the pooled degrees of freedom. This equation is given by

$$\nu = \frac{\left(\frac{s_X^2}{m} + \frac{s_Y^2}{n}\right)^2}{\frac{\left(\frac{s_X^2}{m}\right)^2}{m-1} + \frac{\left(\frac{s_Y^2}{n}\right)^2}{n-1}} \, ,$$

and is commonly used in scenarios when two population variances are unknown, and there is no evidence to assume that they are equal.

Hence, in this situation, one may construct a $100(1-\alpha)\%$ confidence interval for $\mu_X - \mu_Y$ as

$$(\bar{x} - \bar{y}) \pm t_{1-\alpha/2}(\nu)\sqrt{\frac{s_X^2}{m} + \frac{s_Y^2}{n}} \, .$$

Example

The following results were obtained from an experiment that compared the absorption times (in minutes) of two orally administered drugs. The scientist conducting the experiment strongly suspects that the variances of the absorption times of both drugs are not equal.

Drug X:	$m = 14$	$\sum x = 301.6$	$\sum x^2 = 8120.72$
Drug Y:	$n = 10$	$\sum y = 210.0$	$\sum y^2 = 5207.49$

Find a 95% confidence interval for the difference between the mean absorption times of the two drugs.

Solution: The 95% confidence interval for difference between the mean absorption times of the two drugs is given by

$$(\bar{x} - \bar{y}) \pm t_{0.975}(v)\sqrt{\frac{s_X^2}{m} + \frac{s_Y^2}{n}}.$$

The sample variances for each drug are given by

$$s_X^2 = \frac{1}{m-1}\left\{\sum_{i=1}^{m} x_i^2 - \frac{1}{m}\left(\sum_{i=1}^{m} x_i\right)^2\right\} = \frac{1}{13}\left\{8120.72 - \frac{1}{14}(301.6)^2\right\} = 124.8764835$$

$$s_Y^2 = \frac{1}{n-1}\left\{\sum_{i=1}^{n} y_i^2 - \frac{1}{n}\left(\sum_{i=1}^{n} y_i\right)^2\right\} = \frac{1}{9}\left\{5207.49 - \frac{1}{10}(210.0)^2\right\} = 88.61$$

and Welch–Satterthwaite's approximation is given by

$$v = \frac{\left(\frac{s_X^2}{m} + \frac{s_Y^2}{n}\right)^2}{\frac{\left(\frac{s_X^2}{m}\right)^2}{m-1} + \frac{\left(\frac{s_Y^2}{n}\right)^2}{n-1}} = \frac{\left(\frac{124.8764835}{14} + \frac{88.61}{10}\right)^2}{\frac{\left(\frac{124.8764835}{14}\right)^2}{13} + \frac{\left(\frac{88.61}{10}\right)^2}{9}} = 21.29808397 \approx 21.$$

The 95% confidence interval for difference between the mean absorption times of the two drugs is therefore

$$(\bar{x} - \bar{y}) \pm t_{0.975}(v)\sqrt{\frac{s_X^2}{m} + \frac{s_Y^2}{n}} = (21.54285714 - 21.0) \pm t_{0.975}(21)\sqrt{\frac{124.8764835}{14} + \frac{88.61}{10}}$$

$$= [-8.2279, 9.3136],$$

to four decimal places where $t_{0.975}(21) = 2.080$.

◀

4.2.2   Confidence Intervals for Unknown Variances

4.2.2.1 Confidence Interval for a Variance
First recall Lemma 2.2, which is repeated below:

Let $X_1, X_2, \ldots, X_n$ be a random sample from a Normal distribution with mean μ, variance σ^2 and sample variance S^2. Then $\chi^2 = \dfrac{(n-1)S^2}{\sigma^2}$ has a chi-squared distribution with $(n-1)$ degrees of freedom. Briefly, we write $\chi^2 \sim \chi^2(n-1)$.

The idea is to find two points on the chi-squared distribution, such that the probability that the random variable χ^2 lies in between these two points is $(1 - \alpha)$. This is demonstrated in the picture below.

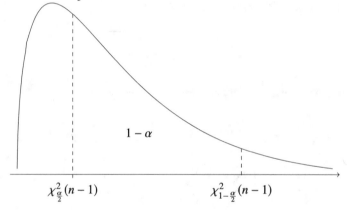

There are many vertical dashed lines we could choose so that the amount of area under the curve is $(1 - \alpha)$. One choice for the placement of these dashed lines is at $\chi^2_{\frac{\alpha}{2}}(n-1)$ and $\chi^2_{1-\frac{\alpha}{2}}(n-1)$, respectively. It follows from the definition of these quantities that

$$P\left[\chi^2_{\frac{\alpha}{2}}(n-1) < \chi^2 < \chi^2_{1-\frac{\alpha}{2}}(n-1)\right] = (1 - \alpha).$$

Recall that $\chi^2_{\frac{\alpha}{2}}(n-1)$ is the value such that the probability that a chi-squared random variable with $(n-1)$ degrees of freedom being smaller than it is $\alpha/2$.
We can write the following:

$$P\left[\chi^2_{\frac{\alpha}{2}}(n-1) < \frac{(n-1)S^2}{\sigma^2} < \chi^2_{1-\frac{\alpha}{2}}(n-1)\right] = P\left[\frac{(n-1)S^2}{\chi^2_{1-\frac{\alpha}{2}}(n-1)} < \sigma^2 < \frac{(n-1)S^2}{\chi^2_{\frac{\alpha}{2}}(n-1)}\right]$$

$$= (1 - \alpha)$$

and a $100(1 - \alpha)\%$ confidence interval for σ^2 is given by

$$
\left[\frac{(n - 1)s^2}{\chi^2_{1-\frac{\alpha}{2}}(n - 1)}, \frac{(n - 1)s^2}{\chi^2_{\frac{\alpha}{2}}(n - 1)} \right].
$$

Example

In 16 experiments, an ice-cream dispensing machine was found to have a sample variance of 4 mL. Find a 99% confidence interval for the population variance.

Solution: Assume that the data is a random sample from a Normally distributed population. From statistical tables $\chi^2_{\frac{\alpha}{2}}(n - 1) = \chi^2_{0.005}(15) = 4.601$ and $\chi^2_{1-\frac{\alpha}{2}}(n - 1) = \chi^2_{0.995}(15) = 32.801$. A 99% confidence interval for the population variance is given by

$$
\left[\frac{(n - 1)s^2}{\chi^2_{1-\frac{\alpha}{2}}(n - 1)}, \frac{(n - 1)s^2}{\chi^2_{\frac{\alpha}{2}}(n - 1)} \right] = \left[\frac{15 \times 4}{32.801}, \frac{15 \times 4}{4.601} \right] = [1.8292, 13.0406].
$$

◄

R Example

We now consider how to write functions in R that will allow us to obtain some of the confidence intervals we have considered. We will develop code for confidence intervals for (i) an unknown mean with random sample taken from a Normally distributed population and unknown population variance and (ii) for an unknown population variance. We directly create a function which replicates the formula for the confidence interval for each, which have already been derived. Firstly, for the unknown mean we can construct the following:

```
mean.ci <- function(data, conf.level = 0.95){
  xbar = mean(data)
  sx   = sd(data)
  df   = length(data)-1
  t    = qt((1-conf.level)/2, df, lower.tail = FALSE)
  c(xbar-t*sx/sqrt(length(data)),xbar+t*sx/sqrt(length(data)))
}
```

and for the unknown variance we can construct the following:

```
var.ci = function(data, conf.level = 0.95) {
  df = length(data)-1
  chilower = qchisq((1-conf.level)/2, df)
  chiupper = qchisq((1-conf.level)/2, df, lower.tail = FALSE)
  v = var(data)
  c(df * v/chiupper, df * v/chilower)
}
```

Suppose the following data of length of five mice in centimeters were obtained:

```
mice <- c(8.8, 8.9, 8.7, 8.4, 9.0)
```

then the commands

```
mean.ci(mice)
var.ci(mice)
```

find the 95% confidence intervals for the unknown mean and variance to be

```
8.474147   9.045853
0.01902491 0.43763807
```

As another example to illustrate the meaning of confidence intervals, let us suppose that we have a Normally distributed population, with mean 0 and variance 1. We take 100 samples of size 50 from this population, and for each sample find the confidence interval for the unknown population mean using the function mean.ci above.

```
samples <- matrix(rnorm(100 * 50), 50)
for (i in 1:100){
  conf.int <- apply(samples, 2, mean.ci)
}
```

We can now plot the confidence intervals as horizontal lines with a vertical line at the population mean of 0 using the code

```
plot(range(conf.int), c(0, 101), type = "n",
    xlab = "Confidence Interval", ylab = "Sample Number")
for (i in 1:100) lines(conf.int[,i], rep(i, 2), lwd = 2,
    lty = ifelse((conf.int[1,i]<0 & conf.int[2,i]<0) |
    (conf.int[1,i]>0 & conf.int[2,i]>0),2,1))
abline(v = 0)
```

and obtain the following:

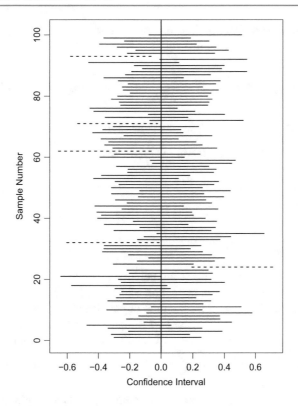

where the horizontal line denoting the confidence interval computed from each of our samples is bold if it contains the population mean and is dashed if it doesn't.

We now recall our early discussion of confidence intervals. In the long run, 95% of confidence intervals will contain $\mu = 0$. In the figure above, we can see this is the case. There are 5 out of the 100 generated confidence intervals which do not include the true population mean $\mu = 0$, but 95 out of the 100 generated confidence intervals do contain $\mu = 0$.

◀

4.2.2.2 Confidence Interval for a Ratio of Two Variances

First recall Lemma 2.3 which is repeated below:

Let $X_1, X_2, \ldots, X_m$ and $Y_1, Y_2, \ldots, Y_n$ be random samples from $X \sim N[\mu_X, \sigma_X^2]$ and $Y \sim N[\mu_Y, \sigma_Y^2]$ respectively. Let S_X^2 and S_Y^2 be the corresponding sample variances. It follows that $\dfrac{(m-1)S_X^2}{\sigma_X^2} \sim \chi^2(m-1)$ and $\dfrac{(n-1)S_Y^2}{\sigma_Y^2} \sim \chi^2(n-1)$. Hence

$$\frac{S_X^2/\sigma_X^2}{S_Y^2/\sigma_Y^2} \sim F(m-1, n-1).$$

Hence, there exists two points on the F distribution, $F_{\alpha/2}(m-1, n-1)$ and $F_{1-\alpha/2}(m-1, n-1)$ such that

$$P\left[F_{\alpha/2}(m-1, n-1) < \frac{S_X^2/\sigma_X^2}{S_Y^2/\sigma_Y^2} < F_{1-\alpha/2}(m-1, n-1)\right] = 1 - \alpha.$$

Rearranging the inequality in the square brackets, one may derive that the $100(1-\alpha)\%$ confidence interval for $\frac{\sigma_X^2}{\sigma_Y^2}$ is given by

$$\left[\frac{s_X^2}{s_Y^2}\frac{1}{F_{1-\alpha/2}(m-1, n-1)}, \frac{s_X^2}{s_Y^2}\frac{1}{F_{\alpha/2}(m-1, n-1)}\right].$$

Example

A new car was tested 10 times on fuel X and 5 times on fuel Y, with the following sample variances in miles per gallon achieved: $s_X^2 = 20$ and $s_Y^2 = 35.6$. Obtain a 95% confidence interval for the ratio of population variances.

Solution: Assume that the two samples of data are random samples from independent, Normally distributed populations. From statistical tables $F_{\alpha/2}(m - 1, n - 1) = F_{0.025}(9, 4) = \dfrac{1}{F_{0.975}(4, 9)} = \dfrac{1}{4.718} \approx 0.212$ and $F_{1-\alpha/2}(m - 1, n - 1) = F_{0.975}(9, 4) = 8.905$.

A 95% confidence interval for the ratio of population variances is given by

$$\left[\frac{s_X^2}{s_Y^2}\frac{1}{F_{1-\alpha/2}(m-1, n-1)}, \frac{s_X^2}{s_Y^2}\frac{1}{F_{\alpha/2}(m-1, n-1)}\right] = \left[\frac{20}{35.6}\times\frac{1}{8.905}, \frac{20}{35.6}\times\frac{1}{0.212}\right]$$

$$= [0.063, 2.650].$$

◄

4.3 Exercises

1. A random sample of 25 observations from a Normal distribution having known variance 25 has a sample mean of 80. A random sample of 25 observations from another (independent) Normal distribution with known variance 25 had a sample mean of 76. Calculate

   a. 95% confidence intervals for each of the population means.
   b. A 95% confidence interval for the difference between the population means.

2. Six biscuits in a factory are chosen at random, and the amount of fat in them is measured. The results (in grams) were as follows: 0.9, 1.0, 1.0, 1.0, 1.2, 1.2. Clearly stating your assumptions, compute 95% confidence intervals for

a. the population mean,
b. the population variance.

3. A random sample from a Normal distribution with unknown mean μ and variance 16 was used to obtain the 99% confidence interval $(27.7533, 30.1327)$ for μ. Deduce the value of (a) the sample mean and (b) the sample size.
4. The following results were obtained from an experiment that compared the absorption times (in minutes) of two orally administered drugs. The scientist conducting the experiment strongly suspects that the variances of the absorption times of both drugs are not equal.

Drug X:	$m = 14$	$\sum x = 301.6$	$\sum x^2 = 8120.72$
Drug Y:	$n = 10$	$\sum y = 210.0$	$\sum y^2 = 5207.49$

Throughout this question, state clearly any assumptions that you make.

a. Find a 95% confidence interval for the variance of the absorption time of drug X.
b. Find a 95% confidence interval for the variance of the absorption time of drug Y.
c. Find a 95% confidence interval for the ratio of the variances of the absorption times of the two drugs, and interpret your findings with particular relation to the scientists hypothesis regarding the variances. You may assume that $F_{0.975}(13, 9) = 3.831$ and $F_{0.025}(13, 9) = 0.302$.

5. In five attempts, a girl completed a Rubix cube in

$$135.4, 152.1, 146.7, 143.5, 146.0$$

seconds. In different five attempts, a boy completed the cube in

$$133.1, 126.9, 129.0, 139.6, 144.0$$

seconds. Find a 90% confidence interval for the difference in mean time taken to complete the cube assuming that the both samples are from independent Normal populations with equal variances. Also, find a 95% confidence interval for the ratio of population variances.

Hypothesis Testing

5

5.1 Introduction

Suppose we wish to test a claim. Examples could be the following:

- Is a coin fair?
- Is the weather more variable in summer or winter?
- Is a new drug better than an existing one for treating an illness?
- Would an alteration to a car engine yield it more fuel efficient?

In statistical inference, using data to assess the truth (or otherwise) of a claim is known as hypothesis testing.

A statistical hypothesis is a statement hypothesizing a value for an unknown population parameter. For example, consider the hypothesis that a coin is fair. Tossing coins can be modeled as a Binomial random variable, where we toss the coin n times and have a probability p of obtaining a head. The hypothesis that a coin is fair is equivalent to testing the hypothesis that the parameter $p = \frac{1}{2}$ or not. We can take a random sample of coin tosses and attempt to assess the evidence for or against this hypothesis.

Hypothesis testing has lots of terminology. As far as possible we will remain general and test hypotheses concerning an unknown population parameter θ. The purpose of the hypothesis test is to choose between two hypotheses. One is called the null hypothesis, and the other is called the alternative hypothesis. The null hypothesis is denoted H_0, and the alternative hypothesis is denoted H_1.

The null hypothesis is commonly of the form

$$H_0 : \theta = \theta_0 ,$$

where H_0 is the notation to define the null hypothesis, θ is an unknown population parameter and θ_0 is some numerical value.

© Springer Nature Switzerland AG 2020
J. Gillard, *A First Course in Statistical Inference*,
Springer Undergraduate Mathematics Series,
https://doi.org/10.1007/978-3-030-39561-2_5

Corresponding to this null hypothesis, the alternative hypothesis H_1 is one of the following:

$$H_1 : \theta \neq \theta_0 \,,$$
$$H_1 : \theta < \theta_0 \,,$$
$$H_1 : \theta > \theta_0 \,.$$

The first option is known as "two-tailed", while the latter two are known as "one-tailed".

In this framework, we have two possibilities:

1. We can reject H_0, and thus accept H_1, or
2. We can accept H_0, and thus reject H_1

and consequently there are two types of errors that we can make:

1. The incorrect rejection of H_0, known as a Type I error, which we say happens with probability α,
2. The incorrect rejection of H_1, known as a Type II error, which we say happens with probability β.

The probability of making a Type I error is denoted α, and this is known as the *significance level* of the hypothesis test. The probability of making a Type II error is denoted β. Another concept used in hypothesis testing is the so-called *power* of the hypothesis test, and this is given by $1 - \beta$. Naturally, for any hypothesis test, we desire α to be as small as possible, and the power $1 - \beta$ to be as large as possible.

We now address the problem of how to make the decision of accepting or rejecting H_0. This is decided on the basis of a *test statistic*. We will consider an example to illuminate some of these concepts described so far, and also outline each of the main steps necessary.

Example

A beer-dispensing machine is supposed to deliver 20 fluid ounces of beer. The amount dispensed by the machine is thought to be Normally distributed. Eight samples are measured from the machine, with the following results: 20.89, 20.90, 20.87, 20.94, 20.92, 20.90, 20.93, and 20.93. The sample mean is given by $\bar{x} = 20.91$, and the sample standard deviation is given by $s = 0.023904$. Test the hypothesis that the mean amount dispensed by the machine is indeed 20 fluid ounces.

Solution: We begin by forming our hypothesis that we wish to test. The hypothesis is concerning a mean amount dispensed by the machine, and a population mean is typically denoted by μ. Our null hypothesis is thus

$$H_0 : \mu = 20,$$

and we now have to form our alternative hypothesis. For this, we look at the context of our problem: do we care if the machine dispenses more than 20 fluid ounces? Do we care if the machine dispenses less than 20 fluid ounces? This isn't clear. We will proceed with a two-tailed alternative hypothesis as there is no steer to the other possible versions of the alternative. Our alternative hypothesis is thus

$$H_1 : \mu \neq 20.$$

Let us seek for some evidence for the truth (or otherwise) of the null hypothesis H_0. We have some data, namely, a sample mean and a sample variance. If we assume that our observed sample is independent and identically distributed from a Normal population, then the random variable $T = \dfrac{\bar{X} - \mu}{S/\sqrt{n}}$ follows a Student's t-distribution with $(n - 1)$ degrees of freedom. That is, $T \sim t(n - 1)$.

Using our sample mean and variance, we can substitute the random variables in T, namely, $\bar{X}$ and S, with numerical values computed from the observed sample, $\bar{x} = 20.91$ and $s = 0.023904$, respectively, giving

$$t = \frac{20.91 - \mu}{0.023904/\sqrt{8}} .$$

The value of t can be viewed as an element sampled from the Student's t-distribution with $8 - 1 = 7$ degrees of freedom. We are unable to "compute it" at the moment as we do not have a value for μ.

What follows is the fundamental step of hypothesis testing. We proceed by assuming that the null hypothesis is true. This would give us a value of μ (since if H_0 is true then $\mu = 20$) that would lead us to be able to compute t. As we know the distribution from which t is sampled, we can then work out how likely t is to happen, which is a proxy for asking how likely $\mu = 20$ is to happen as this is the very thing we needed to assume true in order to compute t in the first place.

Assuming the null hypothesis true, then

$$t = \frac{20.91 - \mu}{0.023904/\sqrt{8}} = \frac{20.91 - 20}{0.023904/\sqrt{8}} \approx 107.68 .$$

The above is known as the *test statistic*. We evaluate how likely it is to happen in an attempt to answer how likely our null hypothesis is to happen.

There are two (related) approaches as to how to proceed from here. One is based on constructing an *acceptance region* from a significance level (α, which is the probability of a Type I error) that is fixed before conducting the hypothesis test. The acceptance region is a range of values of the test statistic which lead to the acceptance of the null hypothesis. The second approach is based on the computation of a so-called *p-value* which is a more detailed description of where our value of the test statistic t lies on the Student's t-distribution. Test statistics which lie in the tails of the distribution are unlikely to happen, which means the assumption we needed to make to compute it (that the null hypothesis is true) is also unlikely to happen. We will take each approach in turn.

Approach 1: Constructing an Acceptance Region From a Given Significance Level α

We wish to construct a region of test statistics that would lead to the acceptance of the null hypothesis, and we have a specified significance level α set in advance. Recall that α is the probability of making a Type I error. Typically α is taken to be 5%.

An acceptance region $\mathcal{A}$ could look like $\mathcal{A} = \{t : -k \leq t \leq k\}$ for some value of k. We also introduce the idea of a critical region C which is the region of test statistics that would lead to the rejection of the null hypothesis. For the acceptance region $\mathcal{A}$ given, the critical region would be $C = \{t : t < -k \text{ or } t > k\} = \{t : |t| > k\}$. We wish our test statistic to be inside the acceptance region $\mathcal{A}$ if the null hypothesis H_0 is true, and wish it to be inside the critical region C if the null hypothesis H_0 is false.

If our test statistic is inside the critical region when the null hypothesis is true, then we have made a Type I error, which happens with probability α. We can use this fact to deduce what the value of k should be, since

$$
\begin{aligned}
\alpha = P[C \mid \mu = 20] &= 1 - P[\mathcal{A} \mid \mu=20] = 1 - P[-k \leq T \leq k \mid \mu = 20] \\
&= 1 - (P[T \leq k] - P[T \leq -k]) \\
&= 1 - (P[T \leq k] - \{1 - P[T \leq k]\}) \\
&= 2(1 - P[T \leq k])
\end{aligned}
$$

and so solving this equation, k should be chosen such that $P[T \leq k] = 1 - \frac{\alpha}{2}$. Recall that the value t such that $P[T \leq t] = \gamma$ is given by $t_\gamma (n-1)$ and so here $k = t_{1-\frac{\alpha}{2}} (n-1)$. Such values can be found from Table C.4.

For a 5% significance level and our example with the beer-dispensing machine

$$
\mathcal{A} = \left\{ t : -t_{1-\frac{\alpha}{2}} (n-1) \leq t \leq t_{1-\frac{\alpha}{2}} (n-1) \right\} = \{t : -t_{0.975}(7) \leq t \leq t_{0.975}(7)\}
$$

$$
= \{t : -2.365 \leq t \leq 2.365\}
$$

and our value of the test statistic $t = 107.68$ does not lie in this region (so it lies in the critical region C). We therefore reject the null hypothesis that the population mean is 20 with significance level $\alpha = 5\%$. Note that this derivation of the acceptance region $\mathcal{A}$ is for the two-tailed alternative hypothesis $H_1 : \mu \neq 20$. We now proceed with investigating the form of the acceptance regions for the two possible one-tailed alternative hypotheses: (i) $H_1 : \mu < 20$ or (ii) $H_1 : \mu > 20$. For these one-tailed hypotheses, the acceptance regions are $\mathcal{A} = \{t : -k \leq t\}$ or $\mathcal{A} = \{t : t \leq k\}$ with critical regions $C = \{t : t < -k\}$ or $C = \{t : t > k\}$, respectively. Again we can use the fact that $\alpha = P[C \mid \mu = 20]$ to deduce the value of k in these one-tailed cases.

For $\mathcal{A} = \{t : -k \leq t\}$,

$$\alpha = P[C \mid \mu{=}20] = 1 - P[\mathcal{A} \mid \mu = 20] = 1 - P[-k \leq T] = 1 - P[T \leq k],$$

so we choose the value of k such that $P[T \leq k] = 1 - \alpha$, yielding $k = t_{1-\alpha}(n-1)$ and the acceptance region $\mathcal{A} = \{t : -t_{1-\alpha}(n-1) \leq t\}$.
For $\mathcal{A} = \{t : t \leq k\}$,

$$\alpha = P[C \mid \mu = 20] = 1 - P[\mathcal{A} \mid \mu = 20] = 1 - P[T \leq k],$$

so we choose the value of k such that $P[T \leq k] = 1 - \alpha$, yielding $k = t_{1-\alpha}(n-1)$ and the acceptance region $\mathcal{A} = \{t : t \leq t_{1-\alpha}(n-1)\}$.
For a 5% significance level these acceptance regions become $\mathcal{A} = \{t : -1.895 \leq t\}$ and $\mathcal{A} = \{t : t \leq 1.895\}$, respectively, for our example where $n = 8$.

Approach 2: Computing a p-value

In the previous approach, it is clear that our test statistic is "well outside" the critical region. The width of the acceptance region $\mathcal{A}$ is dictated by α. Below is a table of the acceptance regions for different α for our example (where $n = 8$) with a two-tailed alternative hypothesis:

α	$\mathcal{A}$
5%	$\{t : -2.365 \leq t \leq 2.365\}$
2%	$\{t : -2.998 \leq t \leq 2.998\}$
1%	$\{t : -3.499 \leq t \leq 3.499\}$

So for $\alpha = 1\%$ the test statistic still lies outside the acceptance region. It is desirable that α is as small as possible, as this is the probability of making an error. If we fix a significance level in advance, then our decision becomes binary: our test statistic either lies in the acceptance region or not. A lot of information is lost this way. As the above table shows, the same conclusion would have been reached for a smaller α.

The approach using so-called p-values tries to circumnavigate the above by giving a better idea of where the test statistic lies on the distribution. We begin by defining what p-values are.

Definition 5.1 A p-value is the probability of obtaining a test statistic, or more extreme, if the null hypothesis is true.

What we mean by "extreme" is clarified by revisiting our example. A more extreme, or unlikely test statistic would be to observe something larger than 107.68. But as we have a two-tailed alternative we have to take into account another possible extreme that the test statistic could be smaller than -107.68. The p-value p for this example is thus

$$p = P[T > 107.68] + P[T < -107.68] = 2P[T > 107.68].$$

Suppose instead we had a one-tailed alternative hypothesis. In this case, the p-value is $p = P[T > 107.68]$ which by symmetry of the Student's t-distribution is the same as $P[T < -107.68]$. For all test statistics derived from symmetric distributions (e.g., Normal or Student's t-distributions), the p-value for a one-tailed test is half that of a two-tailed test.

To approximate the p-value p for our two-tailed test, we note that $t_{0.995}(7) = 3.499$. This means that $P[T < 3.499] = 0.995$ (shaded in gray in the below diagram) or $P[T > 3.499] = 0.005$. Our value of the test statistic, $t = 107.68$, is much further to the right than 3.499 and so $P[T > 107.68] << 0.005$ (shaded in black) and $p = 2P[T > 107.68] << 0.01$. The p-value for a one-tailed hypothesis test is $p = P[T > 107.68] = P[T < -107.68] << 0.005$.

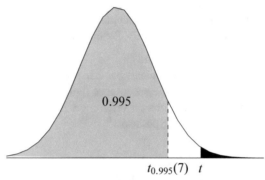

$$t_{0.995}(7) \quad t$$

Small p-values could suggest more evidence to reject the null hypothesis. Traditionally interpretations of a p-value follow guidelines such as those listed below:

- $p < 0.01$, there is very strong evidence for rejecting H_0;
- $0.01 \leq p \leq 0.05$, there is strong evidence for rejecting H_0;
- $p > 0.05$, there is insufficient evidence for rejecting H_0;

but the use (and abuse) of p-values is something which is richly debated in the statistical literature.

Using these guidelines for our example, there is strong evidence for rejecting H_0.

We now describe another example using the Student's t-distribution to test a hypothesis about an unknown population mean. Such hypothesis tests are classically called *t-tests* and are widely used in practice.

Example

It is claimed that a shop makes a profit of £850, per week, on average. To test this, the manager records the weekly profit across five randomly selected weeks, and finds the average to be £905, with standard deviation £50. The manager asks the question "Have profits increased significantly?". Perform a suitable hypothesis test at a 5% significance level and also approximate the p-value.

Solution: Let μ denote the population mean. The hypotheses are $H_0 : \mu = 850$ and $H_1 : \mu > 850$. If we assume that our observed sample is independent and identically distributed from a Normal population, then the random variable

$$T = \frac{\bar{X} - \mu}{S/\sqrt{n}} \sim t(n-1).$$

Assuming that the null hypothesis H_0 is true, our test statistic is $t = \dfrac{905 - 850}{50/\sqrt{5}} \approx$ 2.4597. For a 5% significance level with a one-tailed alternative hypothesis, our acceptance region is

$$\mathcal{A} = \{t : t \le t_{1-\alpha}(n-1)\} = \{t : t \le t_{0.95}(4)\} = \{t : t \le 2.132\}$$

and our value of the test statistic $t = 2.4597$ does not lie in this region. We reject the null hypothesis that the population mean is £850.

The p-value is $p = P[T > 2.4597]$. From statistical tables $t_{0.95}(4) = 2.132$ (meaning $P[T > 2.132] = 0.05$) and $t_{0.975}(4) = 2.776$ (meaning $P[T > 2.776] = 0.025$) hence $0.025 < P[T > 2.4597] < 0.05$. It follows that the p-value p obeys the inequality $0.025 < p < 0.05$.

◀

R Example

We will study the iris data set inbuilt in R which gives the measurements in centimeters of the sepal length, sepal width, petal length, and petal width, for 50 flowers from three species of iris. The species are iris setosa, versicolor, and virginica.

We load and inspect the top of the data by running the following:

```
data("iris")
head(iris)
```

which produces the following output:

	Sepal.Length	Sepal.Width	Petal.Length	Petal.Width	Species
1	5.10	3.50	1.40	0.20	setosa
2	4.90	3.00	1.40	0.20	setosa
3	4.70	3.20	1.30	0.20	setosa
4	4.60	3.10	1.50	0.20	setosa
5	5.00	3.60	1.40	0.20	setosa
6	5.40	3.90	1.70	0.40	setosa

Suppose we wish to test the hypothesis that the average sepal length is 6 against the two-tailed alternative. We do this by invoking the following:

```
attach(iris)
t.test(Sepal.Length, mu=6)
```

which obtains the following output:

```
        One Sample t-test

data:  Sepal.Length
t = -2.3172, df = 149, p-value = 0.02186
alternative hypothesis: true mean is not equal to 6
95 percent confidence interval:
 5.709732 5.976934
sample estimates:
mean of x
 5.843333
```

The output can be seen to contain the value of the test statistic, the numbers of degrees of freedom (df) and the p-value of the test which suggests there is strong evidence for rejecting the null hypothesis that the true mean is equal to 6 cm. Other information such as the 95% confidence interval for the mean is also provided.

◀

All examples thus far have concerned hypothesis testing using the Student's t-distribution, the so-called t-tests. We will now describe how to use the chi-squared and F-distributions to perform hypothesis tests regarding population variance(s).

If we wish to test a hypothesis concerning a single variance, i.e.,

$$H_0 : \sigma^2 = \sigma_0^2, \qquad H_1 : \sigma^2 \neq \sigma_0^2,$$

where σ^2 is an unknown population variance and σ_0^2 is some given hypothesized numerical value then an appropriate distribution (assuming a random sample has been taken) is the chi-squared distribution since

$$\chi^2 = \frac{(n-1)S^2}{\sigma^2} \sim \chi^2(n-1).$$

This distribution takes values on $(0, \infty)$ and isn't symmetric, which results in some technical and operational differences from t-tests. The shape of the distribution (a typical example of the probability density function of the chi-squared distribution is reproduced below) is skewed to the left.

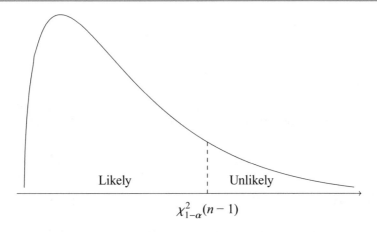

$$\chi^2_{1-\alpha}(n-1)$$

Unlikely events are thus concentrated toward the right-hand tail of the distribution. Likely events are concentrated at the left-hand side of the distribution. So it makes sense, upon assuming the null hypothesis H_0 true, to have an acceptance region of the form

$$\mathcal{A} = \left\{\chi^2 : 0 \le \chi^2 \le \chi^2_{1-\alpha}(n-1)\right\}$$

and thus a critical region of the form

$$\mathcal{C} = \left\{\chi^2 : \chi^2 > \chi^2_{1-\alpha}(n-1)\right\}$$

for a significance level α.

If we wish to test a hypothesis concerning two variances, i.e.,

$$H_0 : \sigma_X^2 = \sigma_Y^2, \qquad H_1 : \sigma_X^2 \ne \sigma_Y^2,$$

where σ_X^2 and σ_Y^2 are two unknown population variances, then an appropriate distribution would be the F-distribution since

$$F = \frac{S_X^2/\sigma_X^2}{S_Y^2/\sigma_Y^2} \sim F(m-1, n-1).$$

Here S_X^2 and S_Y^2 are sample variances of two independent random samples of sizes m and n, respectively. Assuming the null hypothesis H_0 true, then the test statistic f becomes

$$f = \frac{s_x^2}{s_y^2},$$

and analogous to the discussion regarding the shape of the chi-squared distribution, the F-distribution is also skewed to the left. For this hypothesis test with a significance level of α, the acceptance region $\mathcal{A}$ is given by

$$\mathcal{A} = \left\{f : 0 \le f \le F_{1-\alpha}(m-1, n-1)\right\},$$

and critical region

$$C = \{f : f > F_{1-\alpha}(m - 1, n - 1)\} .$$

Example

Five sheep were fed on diet A and another six sheep were fed on diet B. The following results for the weight increase in kilograms were obtained (in obvious notation):

$$\bar{x}_A = 7.99, \ \bar{x}_B = 7.13$$
$$s_A^2 = 1.20, \ s_B^2 = 1.00$$

Using a 5% significance level:

1. Test the hypothesis $H_0 : \sigma_A^2 = \sigma_B^2$ against the alternative $H_1 : \sigma_A^2 \neq \sigma_B^2$ where σ_A^2, σ_B^2 are the population variances for the A and B diets, respectively.
2. Using your results from (1.) test the hypothesis $H_0 : \mu_A = \mu_B$ against the alternative $H_1 : \mu_A > \mu_B$, where μ_A and μ_B are the population mean weight increases for the A and B diets, respectively. Give the p-value for your test.

Solution:

1. Recall the following from Lemma 2.3:

Let $X_1, X_2, \ldots, X_m$ and $Y_1, Y_2, \ldots, Y_n$ be random samples from $X \sim N[\mu_X, \sigma_X^2]$ and $Y \sim N[\mu_Y, \sigma_Y^2]$ respectively. Let S_X^2 and S_Y^2 be the corresponding sample variances. It follows that $\dfrac{(m-1)S_X^2}{\sigma_X^2} \sim \chi^2(m-1)$ and $\dfrac{(n-1)S_Y^2}{\sigma_Y^2} \sim \chi^2(n-1)$. Hence

$$\frac{S_X^2/\sigma_X^2}{S_Y^2/\sigma_Y^2} \sim F(m-1, n-1).$$

Assuming that our data follows the assumptions as detailed in Lemma 2.3, under the null hypothesis H_0 a suitable test statistic can be computed as

$$f = \frac{s_A^2/\sigma_A^2}{s_B^2/\sigma_B^2} = \frac{s_A^2}{s_B^2} = \frac{1.20}{1.00} = 1.20$$

For a 5% significance level, the acceptance region $\mathcal{A}$ is given by

$$\mathcal{A} = \{f : 0 \leq f \leq F_{1-\alpha}(m-1, n-1)\} = \{f : 0 \leq f \leq F_{0.95}(4, 5)\}$$
$$= \{f : 0 \leq f \leq 5.192\} .$$

The test statistic $f = 1.20$ is in the acceptance region $\mathcal{A}$ and so we accept the null hypothesis at a 5% significance level.

2. Assume that both samples are independent, and drawn randomly from Normal populations with equal variances (due to the test performed above this is an appropriate assumption).

The pooled variance estimator of the common variance is

$$s^2 = \frac{(m-1)s_A^2 + (n-1)s_B^2}{m+n-2} = \frac{(4 \times 1.20) + (5 \times 1.00)}{9} \approx 1.0889.$$

The test statistic, assuming the null hypothesis of no difference in the population means, is given by

$$t = \frac{(\bar{x}_A - \bar{x}_B) - (\mu_A - \mu_B)}{\sqrt{s^2 \left(\frac{1}{m} + \frac{1}{n}\right)}} = \frac{7.99 - 7.13}{\sqrt{1.0889 \left(\frac{1}{5} + \frac{1}{6}\right)}} \approx 1.3610.$$

The acceptance region for a 5% significance level is given by

$$\mathcal{A} = \{t : t \leq t_{1-\alpha}(m+n-2)\} = \{t : t \leq t_{0.95}(9)\} = \{t : t \leq 1.833\}.$$

The test statistic $t = 1.3610$ is in the acceptance region $\mathcal{A}$ and so we accept the null hypothesis at a 5% significance level. From statistical tables, we can see that $t_{0.9}(9) = 1.383$. Hence $P[T \leq 1.383] = 0.9$ and so $P[T > 1.383] = 0.1$. The p-value can thus be deduced to be $p = P[T > 1.3610] > 0.1$. This suggests that there is little statistical evidence that the mean increase in weights differs for diets A and B.

◀

R Example

Let us again revisit the iris data set considered and described earlier. Recall that we load and inspect the top of the data by running the following:

```
data("iris")
head(iris)
attach(iris)
```

and suppose we wish to test the hypothesis that the species virginica and versicolor share the same population mean, against the alternative that they have different population means. We do this by using the following:

```
t.test(Sepal.Length[Species=="virginica"],
                    Sepal.Length[Species=="versicolor"])
```

which produces the following output:

```
   Welch Two Sample t-test

data:   Sepal.Length[Species == "virginica"] and
              Sepal.Length[Species == "versicolor"]
t = 5.6292, df = 94.025, p-value = 1.866e-07
alternative hypothesis: true difference in means is
             not equal to 0
95 percent confidence interval:
 0.4220269 0.8819731
sample estimates:
mean of x mean of y
    6.588      5.936
```

Welch's two-sample t-test is a hypothesis test between two population means when the two samples have unequal variances. See the discussion corresponding to Eq. (4.1) in the previous chapter. Here, the null hypothesis is overwhelmingly rejected, having a very small p-value, and it seems as though the mean sepal length is different between these two species.

◄

Example

The duration of a "regular" pregnancy (one without medical complications) can be modeled by a Normal distribution with mean 266 days and standard deviation 16 days. In an urban hospital in a relatively deprived area of the UK, a random sample of 61 pregnancies was studied and the duration of each pregnancy determined. The relevant statistics were

$$\sum x = 15568, \quad \sum x^2 = 4054484.$$

Using a 5% significance level:

1. Test whether the variance of this sampled population differs from that of "regular" pregnancy durations.
2. Test whether there is evidence that the mean pregnancy duration for this sampled population differs from 266 days. Find an approximate p-value for your test, and state your conclusions clearly.

Solution:

1. From the data the sample variance s^2 (revisit Definition 2.9) can be computed to be $s^2 = 1355.4372$. Our hypotheses are the following: $H_0 : \sigma^2 = 256$, $H_1 : \sigma^2 \neq 256$, where σ^2 is the population variance.
   We assume to have a random sample from a Normal distribution with mean μ

and variance σ^2. Then the random variable $\chi^2 = \dfrac{(n-1)S^2}{\sigma^2}$ has a chi-squared distribution with $(n-1)$ degrees of freedom.

Assuming the null hypothesis H_0 is true, we use the test statistic

$$\chi^2 = \frac{(n-1)s^2}{\sigma^2} = \frac{60 \times 1355.4372}{256} = 317.68.$$

For a 5% significance level, the acceptance region $\mathcal{A}$ is given by

$$\mathcal{A} = \{\chi^2 : 0 \leq \chi^2 \leq \chi^2_{0.95}(n-1)\} = \{\chi^2 : 0 < \chi^2 < 79.082\}.$$

The test statistic does not lie within the acceptance region and so we reject the null hypothesis.

2. The hypotheses are $H_0 : \mu = 266$, $H_1 : \mu \neq 266$ where μ is the population mean. Under the assumption that the null hypothesis is true, we use the test statistic

$$t = \frac{\bar{x} - \mu}{s/\sqrt{n}} \approx \frac{255.21311 - 266}{\sqrt{1355.4372/61}} \approx -2.2883.$$

For a 5% significance level, the acceptance region is

$$\mathcal{A} = \{t : -t_{0.975}(60) \leq t \leq t_{0.975}(60)\} = \{t : -2.000 \leq t \leq 2.000\},$$

and our value of the test statistic $t = -2.2883$ does not lie in this region (so it lies in the critical region). We hence reject the null hypothesis. From statistical tables, $t_{0.995}(60) = 2.660$ and the p-value is thus given by

$$2P[T > 2.2883] < 0.01.$$

We again assume that the data is a random sample from a Normal distribution as described above.

◄

The hypothesis tests considered so far are based on scenarios when the data are from Normally distributed populations. The Central Limit Theorem can be used to conduct similar tests when the sample sizes are large enough for this theorem to apply. Suppose that X is a random variable with mean μ and variance σ^2. Let $X_1, \ldots, X_n$ be a random sample from the distribution of X, then by the Central Limit Theorem the statistic

$$Z = \frac{\bar{X} - \mu_0}{\sigma/\sqrt{n}}$$

under a null hypothesis $H_0 : \mu = \mu_0$ is approximately $N[0, 1]$ for large n. It follows that this test statistic can be used for hypothesis tests for non-Normally distributed populations.

This method is widely applicable, and the example that will be demonstrated here involves the Binomial distribution. The Binomial distribution is characterized by two parameters: n the number of trials and p the probability of success in each trial. Let's imagine that we want to test hypotheses concerning p. The theory is contained here.

Let Y be the number of successes obtained in n trials and let $\hat{p} = \frac{Y}{n}$ be the observed proportion of successful trials. It follows that Y has a Binomial distribution, $B[n, p]$, which is approximately $N[np, np(1 - p)]$ for large n. Thus, $\hat{p}$ is approximately $N\left[p, \frac{p(1-p)}{n}\right]$ (convince yourself of this by deriving $E[\hat{p}]$ and $Var[\hat{p}]$) and the standardized statistic

$$z = \frac{\hat{p} - p_0}{\sqrt{\frac{p_0(1-p_0)}{n}}},$$

is approximately $N[0, 1]$ under a null hypothesis $H_0 : p = p_0$. We can then accept or reject H_0 by assessing whether the test statistic z lies in the acceptance region or not.

Example

A drug company claims in an advertisement that 60% of people suffering from a certain illness gain instant relief by using a particular product. In a random sample, 106 out of 200 did gain instant relief. Test the validity of the claim within the advert at a 5% significance level.

Solution: Let p be the probability of success. The hypotheses are $H_0 : p = 0.6$, $H_1 : p \neq 0.6$. An estimator of p is $\hat{p} = 106/200 = 0.53$.

Using the Central Limit Theorem and assuming H_0 true, the test statistic is

$$z = \frac{\hat{p} - p_0}{\sqrt{\frac{p_0(1-p_0)}{n}}} = \frac{0.53 - 0.6}{\sqrt{1/200 \times 0.6 \times 0.4}} \approx -2.02.$$

The acceptance region for this test with a 5% significance level is

$$\mathcal{A} = \{z : -z_{0.975} \leq z \leq z_{0.975}\} = \{z : -1.96 \leq z \leq 1.96\}$$

and as $z = -2.02$ lies outside of $\mathcal{A}$ we reject H_0 at a 5% significance level.

◀

5.2   Power

In previous sections, we have designed hypothesis tests considering only the significance level α (the probability of making a Type I error). We haven't considered β, the probability of making a Type II error. We cannot, in general, compute the Type II error probability, denoted β as it varies with the parameter values consistent with H_1. We can, however, regard it as a function of these values and knowledge of this function helps us to assess the performance of the test. We make the following definition.

Definition 5.2 Consider a test of two hypotheses, H_0 and H_1, concerning a parameter θ, with a fixed significance level α. The power function of the test, denoted by $\pi(\theta)$, is the probability of rejecting H_0, and hence accepting H_1, expressed as a function of θ.

Recall that given a significance level α, we accept the null hypothesis if the test statistic lies in the acceptance region $\mathcal{A}$ and reject the null hypothesis if the test statistic lies in the critical region $\mathcal{C}$. The power function is the probability of a statistic being in the critical region expressed as a function of the parameter we are hypothesizing about.

We will use the notation $P[\mathcal{C}; \theta]$ to denote the probability that a statistic lies in the critical region $\mathcal{C}$, with θ as a parameter, and the power function is thus

$$\pi(\theta) = P[\mathcal{C}; \theta] = 1 - P[\mathcal{A}; \theta].$$

The following example shows how the power function can be found and analyzed.

Example

A random variable X is Normally distributed with unknown mean μ and variance 1. Using a random sample of size 100 taken from X and a 5% significance level, compute the power function for the following hypothesis test:

$$H_0 : \mu = 0; \qquad H_1 : \mu \neq 0.$$

Solution: The hypothesis test concerns the unknown population mean denoted μ. The power function is given by $\pi(\mu) = P[\mathcal{C}; \mu] = 1 - P[\mathcal{A}; \mu]$. The acceptance region for this hypothesis test using a 5% significance level is given by

$$\mathcal{A} = \left\{ z : -z_{1-\frac{\alpha}{2}} \leq z \leq z_{1-\frac{\alpha}{2}} \right\} = \{z : -z_{0.975} \leq z \leq z_{0.975}\} = \{z : -1.96 \leq z \leq 1.96\}.$$

Assuming the null hypothesis true, the appropriate test statistic in this case is $z = \dfrac{\bar{x} - \mu}{\sigma/\sqrt{n}} = \dfrac{\bar{x} - 0}{1/\sqrt{100}}$ (with $\mu = 0$ from the null hypothesis, $\sigma^2 = 1$ and $n = 100$) and so we can write the acceptance region $\mathcal{A}$ as

$$\mathcal{A} = \left\{ \bar{x} : -1.96 \leq \frac{\bar{x}}{1/\sqrt{100}} \leq 1.96 \right\} = \left\{ \bar{x} : -1.96\frac{1}{\sqrt{100}} \leq \bar{x} \leq 1.96\frac{1}{\sqrt{100}} \right\}$$

$$= \{\bar{x} : -0.196 \leq \bar{x} \leq 0.196\}.$$

Fig. 5.1 Graph of the power function $\pi(\mu)$

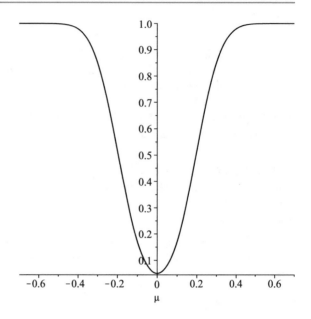

The power function is therefore given by

$$\pi(\mu) = 1 - P\left[-0.196 \le \bar{X} \le 0.196\right] .$$

The dependency of the above on μ is not immediately obvious. Recall that

$$\bar{X} \sim N\left[\mu, \frac{\sigma^2}{n}\right] \sim N\left[\mu, \frac{1}{100}\right]$$

and so

$$\pi(\mu) = 1 - P\left[-0.196 \le \bar{X} \le 0.196\right] = 1 - P\left[\frac{-0.196 - \mu}{1/\sqrt{100}} \le \frac{\bar{X} - \mu}{1/\sqrt{100}} \le \frac{0.196 - \mu}{1/\sqrt{100}}\right]$$

$$= 1 - P\left[\frac{-0.196 - \mu}{1/\sqrt{100}} \le Z \le \frac{0.196 - \mu}{1/\sqrt{100}}\right] ,$$

where Z is a standard Normal random variable, $Z \sim N[0, 1]$.

We can evaluate this power function (to two decimal places) for different values of μ

μ	−0.5	−0.4	−0.3	−0.2	−0.1	0	0.1	0.2	0.3	0.4	0.5
$\pi(\mu)$	1	0.98	0.85	0.52	0.17	0.05	0.17	0.52	0.85	0.98	1

The graph of the power function $\pi(\mu)$ for the above example is given in Fig. 5.1. Three points are worth noting:

1. $\pi(\mu)$ is a probability and lies between 0 and 1.

2. $\pi(0) = P[C; \mu = 0] = \alpha = 0.05$. The logic for this is as follows. If it is the case that $\mu = 0$, then the null hypothesis is indeed true. So in an ideal world the null hypothesis would not be rejected. If the null hypothesis is incorrectly rejected, then this is a Type I error and the probability of making this error is α. Remember that the null hypothesis is rejected if the test statistic lies in the critical region C.
3. As $|\mu|$ increases, $\pi(\mu)$ approaches unity. At the same time, the type II error probability β approaches zero since $\beta = 1 - \pi(\mu)$.

◀

A good hypothesis test will reject H_0 with low probability if it is true but will reject it with high probability if it is false. Thus the power function should be close to zero for values of the parameter consistent with H_0 and close to unity otherwise.

In the above example, the ideal power function is thus

$$\pi(\mu) = \begin{cases} 0, & \text{if } \mu = 0 \\ 1, & \text{if } \mu \neq 0. \end{cases}$$

The ideal power function is not attainable in practice but the closer the actual power function is to it, the better the test. We can usually improve the power function by increasing the sample size.

For example, the table below is the power function of the above example evaluated for different μ, but with a much larger sample size of $n = 500$. Compare this with the table for when $n = 100$. The hypothesis test is more likely to yield the correct decision if n is larger.

μ	−0.5	−0.4	−0.3	−0.2	−0.1	0	0.1	0.2	0.3	0.4	0.5
$\pi(\mu)$	1	1	1	0.99	0.61	0.05	0.61	0.99	1	1	1

R Example

We can use R to compute the power of some hypothesis tests. Let us suppose we have two samples of observations, and we wish to perform a t-test to test the hypothesis that their population means are equal, against the alternative that they are not. For simplicity, we will assume both samples are from populations with the same population variance $\sigma^2 = 1$.

We can consider power calculations by using the pwr package which can be installed (if not already) using install.packages("pwr"). We load the package using

```
library(pwr)
```

In this example, we are going to consider how power changes with sample size, and something known as effect size. The effect size is the minimum difference we would like to see between the population means, should they be different. This difference is often quantified using so-called Cohen's d, which is given by

$$d = \frac{|\mu_X - \mu_Y|}{\sigma},$$

where μ_X is the population mean of the first group, μ_Y is the population mean of the second group, and σ^2 is the common variance. It is suggested that d values of 0.2, 0.5, and 0.8 represent small, medium, and large effect sizes, respectively.

In this exercise, we will consider the power of the t-test to detect significant differences between the two population means by considering sample sizes going from $n = 1, 2, \ldots, 100$ (in each group, so the total number of observations is $2n$) and effect sizes of 0.2, 0.5, and 0.8 assuming $\sigma^2 = 1$. We will consider a fixed 5% significance level. We can perform this exercise using the following snippet of code:

```
power02<-c()
power05<-c()
power08<-c()

for (i in 1:100){
   power02[i]<-pwr.t.test(n = i, d = 0.2, sig.level = 0.05,
                  type = "two.sample")[["power"]]
   power05[i]<-pwr.t.test(n = i, d = 0.5, sig.level = 0.05,
                  type = "two.sample")[["power"]]
   power08[i]<-pwr.t.test(n = i, d = 0.8, sig.level = 0.05,
                  type = "two.sample")[["power"]]
}
```

and plot the results using

```
par(mfrow=c(1,3))
plot(1:100, power02, main="d = 0.2", ylim=c(0,1),
                  xlab="Sample size", ylab = "Power")
plot(1:100, power05, main="d = 0.5", ylim=c(0,1),
                  xlab="Sample size", ylab = "Power")
plot(1:100, power08, main="d = 0.8", ylim=c(0,1),
                  xlab="Sample size", ylab = "Power")
```

which gives

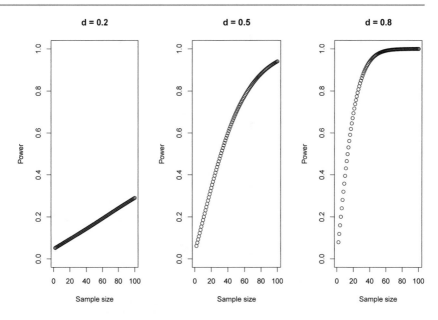

It can be seen that the power is smaller for the smaller effect sizes, and that with $d = 0.2$, a total sample size of 200 gives a weak power probability. For larger effect sizes, stronger power probabilities are observed for smaller sample sizes. The intuition is that big differences are easier to see than subtle differences.

◀

Example

The time to failure of a certain electrical component is known to be distributed with probability density function

$$f(t) = \lambda \exp(-\lambda t), \ t \geq 0.$$

In order to test the null hypothesis $H_0 : \lambda = 2$ against the alternative $H_1 : \lambda > 2$, a random sample of times to failure $T_1, T_2, \ldots, T_n$ is obtained.

Let W denote the sample minimum and Z denote the sample maximum. Consider the test with a critical region of the form $W < k$. Find the value of k in terms of n for which the test has a 5% significance level and obtain an expression for the power function of the test.

Carry out the same exercise for a test whose critical region has the form $Z < k$. Which is the test with higher power? Plot each power function to answer this question with $n = 10$.

Solution: Denote the power function of the test with critical region $W < k$ by $\pi_W(\lambda)$. This power function is given by $\pi_W(\lambda) = P[W < k; \lambda]$. This is an expression for the cumulative distribution function (cdf) of the sample minimum W which is given by $1 - [1 - F(t)]^n$ from Theorem 2.3. The cdf is $F(t) =$

Fig. 5.2 Plot for $\pi_W(\lambda)$ is given in solid line and the plot for $\pi_Z(\lambda)$ is given in dashed line, $n = 10$

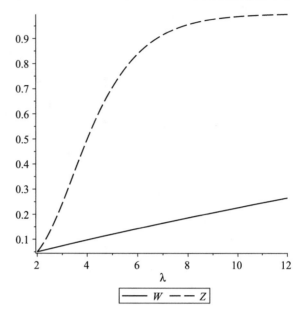

$\int_0^t \lambda \exp(-\lambda t)\, dt = 1 - \exp(-\lambda t)$ and the cdf of the sample minimum W (and hence the power function with critical region $W < k$) is $\pi_W(\lambda) = P[W < k; \lambda] = 1 - \exp(-\lambda n k)$.

For a significance level of 5%, $\pi_W(2) = 1 - \exp(-2nk) = 0.05$ giving $k = -\dfrac{1}{2n}\ln(0.95)$. Substituting this back into the power function, we obtain $\pi_W(\lambda) = 1 - (0.95)^{\lambda/2}$.

Consider now the test with critical region $Z < k$. Let its power function be denoted by $\pi_Z(\lambda) = P[Z < k; \lambda]$. Similarly, this is the cdf of the sample maximum which is given by $[F(t)]^n$ from Theorem 2.3. Hence, $\pi_Z(\lambda) = P[Z < k; \lambda] = (1 - \exp(-\lambda k))^n$ and as before we can use the fact that $\pi_Z(2) = 0.05$ to obtain that for this case $k = -\dfrac{1}{2}\ln(1 - 0.05^{1/n})$ and the power function is $\pi_Z(\lambda) = [1 - (1 - 0.05^{1/n})^{\lambda/2}]^n$.

Plots of both the power functions for $n = 10$ are given in Fig. 5.2. The plot for $\pi_W(\lambda)$ is given in solid line, and the plot for $\pi_Z(\lambda)$ is given in dashed line. The hypothesis test using the sample maximum has higher power.

◀

5.3   Categorical Data

We will now study specific hypothesis tests designed to investigate experiments where the outcome can fall into one of a finite number of categories.

5.3.1 The Chi-Square Goodness-of-Fit Test

Suppose that the result of an experiment is classified into one of k categories $c_1, c_2, \ldots, c_k$. For a given hypothesis the expected numbers of outcomes in each category are $E_1, E_2, \ldots, E_k$ and $O_1, O_2, \ldots, O_k$ are the observed numbers in each category.

As an example, tossing a coin yields a result which could fall into one of the two categories, $c_1 = "Head"$ and $c_2 = "Tail"$. If we toss the coin 50 times and have the hypothesis that the coin is fair, we expect the number of heads and tails to be 25 each, that is, $E_1 = 25$ and $E_2 = 25$. Pragmatically, we know that the results we will observe from tossing a coin 50 times are likely to be different, so we may observe $O_1 = 20$ and $O_2 = 30$ for example. How close the expected values are to the observations will give evidence for or against the hypothesis.

Assuming that the hypothesis which generates the expected values is true, the chi-square statistic is given by

$$\chi^2 = \sum_{i=1}^{k} \frac{(O_i - E_i)^2}{E_i}.$$

The statistic χ^2 has a distribution which is approximately chi-square with $k - m - 1$ degrees of freedom, where m is the number of parameters that need to be estimated from the data to fit any theoretical distribution proposed by the hypothesis. Examples of this will follow.

The approximation is good, provided $E_i > 5$ for all i. If any $E_i < 5$, the outcomes may be grouped to ensure that all $E_i \geq 5$.

It can be seen that χ^2 is a measure of the deviance of the observed numbers O_i seen from our experiment, verses the hypothesized values E_i. The smaller the value of χ^2, the more evidence for the truth of the hypothesis.

The acceptance region for the test assuming the null hypothesis yielding the expected values is given by

$$\mathcal{A} = \{\chi^2 : 0 \leq \chi^2 \leq \chi^2_{1-\alpha}(k - m - 1)\}.$$

Example

A die is rolled 60 times, and the outcome of this experiment is given below. Test at both a 5% and 1% significance levels the null hypothesis H_0 that the die is fair.

Face	1	2	3	4	5	6
Outcome O_i	15	7	4	11	6	17

Solution: If the null hypothesis is true, then we would expect to see, after rolling the die 60 times, 10 of each possible outcome. Now due to experimental variation, it is unlikely given 60 rolls of a die, to perfectly observe 10 ones, 10 twos, etc. So given the data we have above, is it reasonable that the die is fair?

We have already motivated how to calculate the E_i, the expected outcome for each category under the null hypothesis. Thus we can construct the following table:

Face	1	2	3	4	5	6
Outcome O_i	15	7	4	11	6	17
Expected E_i	10	10	10	10	10	10

So we compute the test statistic as

$$\chi^2 = \sum_{i=1}^{6} \frac{(O_i - E_i)^2}{E_i} = 13.6 .$$

We now compare the value of the test statistic with percentage points of the chi-squared distribution, with 5 degrees of freedom

- For a 5% significance level, the acceptance region $\mathcal{A}$ is

$$\mathcal{A} = \{\chi^2 : 0 \le \chi^2 \le \chi^2_{1-\alpha}(k - m - 1)\} = \{\chi^2 : 0 \le \chi^2 \le \chi^2_{0.95}(5)\}$$
$$= \{\chi^2 : 0 \le \chi^2 \le 11.070\}.$$

So at 5% significance, we reject the null hypothesis that the die is fair.
- For a 1% significance level the acceptance region is

$$\mathcal{A} = \{\chi^2 : 0 \le \chi^2 \le \chi^2_{0.99}(5)\} = \{\chi^2 : 0 \le \chi^2 \le 15.086\}.$$

So at 1% significance, we accept the null hypothesis that the die is fair.

Example

100 squares each $1\,\mathrm{m}^2$ were randomly placed in a field where daffodils were growing. The number of clumps of daffodils in each square was counted and the following table of these observed numbers was drawn up.

	0	1	2	3	4	5	≥ 6
O_i	4	17	20	22	15	15	7

A Poisson distribution is suggested as a model for the number of clumps of daffodils in a square. Estimate an appropriate value for the mean of this distribution using the data in the table above. Calculate a table of expected numbers to fit the Poisson distribution to these data and perform a chi-square goodness-of-fit test to determine if the model is a good fit to the observed data at a 10% significance level.

Solution: The total is given by $(0 \times 4) + (1 \times 17) + \ldots + (6 \times 7) = 300$. Hence, an estimate of the mean is $300/100 = 3$.

The formula for the probability density function for a Poisson distribution is

$$f(x; \mu) = \mu^x \exp(-\mu)/x!,$$

where μ is the mean. We need to estimate a value for μ to use this probability density function. We estimate μ by the sample mean, that is, $\hat{\mu} = 3$.

We can then use $f(x; 3)$ to obtain the following table of expected numbers, by multiplying the Poisson probabilities of each possible outcome by 100. Note that the expected number for the last class $x \geq 6$ is found by subtracting the sum of the expected numbers from 0 to 5 inclusive from 100.

	0	1	2	3	4	5	≥ 6
O_i	4	17	20	22	15	15	7
E_i	4.98	14.94	22.40	22.40	16.80	10.08	8.39

For example, the expected number corresponding to 0 clumps of daffodils is given by

$$100 \times f(0; 3) = 100 \times \frac{3^0 \exp(-3)}{0!} = 4.98.$$

Such computations can be performed for 1, 2, 3, 4, or 5 clumps, with the expected number for ≥ 6 found by requiring that the total of the expected numbers should be 100.

From this table, the chi-square goodness-of-fit test statistic is calculated

$$\chi^2 = \sum_{i=1}^{7} \frac{(O_i - E_i)^2}{E_i} = 3.5658.$$

Since one parameter was estimated to derive the expected numbers, the degrees of freedom are 5. There are 7 classes in total, so we calculate the degrees of freedom by taking one away from 7, but then take another one away as we have estimated μ leaving 5.

For a 10% significance level, the acceptance region $\mathcal{A}$ is

$$\mathcal{A} = \{\chi^2 : 0 \leq \chi^2 \leq \chi^2_{1-\alpha}(k - m - 1)\} = \{\chi^2 : 0 \leq \chi^2 \leq \chi^2_{0.90}(5)\}$$
$$= \{\chi^2 : 0 \leq \chi^2 \leq 9.236\}.$$

We thus conclude that there is no significant discrepancy between the fitted numbers and the observed ones, i.e., the Poisson distribution is a good fit to the data, at a 10% significance level.

◄

5.3.2   Contingency Tables for Testing Independence

In the previous example, the probabilities used in the null hypothesis were obtained from a hypothesized model. We often want to test the hypothesis that effects are independent without having a theory to predict the relevant probabilities. In this case, the probabilities must be estimated from a so-called contingency table.

Contingency means dependence, so a contingency table is simply a table that displays how two or more characteristics depend on each other. For example, a contingency table, where Effect 1 has three categories I, II, III and Effect 2 has four categories A, B, C, D, is as follows:

	A	B	C	D	
I	O_{11}	O_{12}	O_{13}	O_{14}	R_1
II	O_{21}	O_{22}	O_{23}	O_{24}	R_2
III	O_{31}	O_{32}	O_{33}	O_{34}	R_3
	C_1	C_2	C_3	C_4	N

where the O_{ij} is the number of observations observed in the ith row and jth column, R_i are row totals, C_j are column totals, and N is the total number of observations.

Under the null hypothesis that the effects are independent, the probability P_{ij} associated with cell (i, j) is estimated by

$$P_{ij} = \frac{R_i}{N} \times \frac{C_j}{N} = \frac{R_i C_j}{N^2}$$

and hence the expected number E_{ij} for each cell (i, j) is given by

$$E_{ij} = N \times P_{ij} = \frac{R_i C_j}{N} .$$

The statistic

$$\chi^2 = \sum_{i=1}^{r} \sum_{j=1}^{c} \frac{(O_{ij} - E_{ij})^2}{E_{ij}} = \sum_{i=1}^{r} \sum_{j=1}^{c} \frac{O_{ij}^2}{E_{ij}} - N,$$

where r is the number of rows, and c is the number of columns follows a chi-squared distribution with $(r - 1)(c - 1)$ degrees of freedom. This can be used to test the hypothesis of independence of two effects. The acceptance region of the test of independence is

$$\mathcal{A} = \left\{ \chi^2 : 0 \leq \chi^2 \leq \chi^2_{1-\alpha}\big((r - 1)(c - 1)\big) \right\},$$

where the null hypothesis is that the effects are independent.

Example

The distribution of five plant species A, B, C, D, E is being investigated at three different locations I, II, III. The contingency table of the results is presented below:

	A	B	C	D	E
I	10	22	38	8	66
II	27	62	120	30	200
III	45	100	207	49	342

Test at a 5% significance level to address whether the species of plant is independent of the location.

Solution: To estimate the probabilities, and thus estimate the expected values, we need to compute row and column totals.

	A	B	C	D	E	
I	10	22	38	8	66	144
II	27	62	120	30	200	439
III	45	100	207	49	342	743
	82	184	365	87	608	1326

Under the null hypothesis of independence (here between plant species and location), we estimate the expected values using the formula given earlier, giving the following table of expected values:

	A	B	C	D	E	
I	8.9	20.0	39.6	9.4	66	144
II	27.1	60.9	120.8	28.8	201.3	439
III	45.9	103.1	204.5	48.7	340.7	743
	82	184	365	87	608	1326

The test statistic is computed as

$$\chi^2 = \sum_{i=1}^{3} \sum_{j=1}^{5} \frac{(O_{ij} - E_{ij})^2}{E_{ij}} = 1.14.$$

The 95th percentage point of the chi-squared distribution with $(r-1)(c-1) = 2 \times 4 = 8$ degrees of freedom is $\chi^2_{0.95}(8) = 15.507$, and so the acceptance region of the hypothesis test is

$$\mathcal{A} = \left\{ \chi^2 : 0 \le \chi^2 \le \chi^2_{0.95}(8) \right\} = \left\{ \chi^2 : 0 \le \chi^2 \le 15.507 \right\}.$$

The null hypothesis that the distribution of the plants is independent of location is accepted at a 5% significance level as the test statistic lies within $\mathcal{A}$.
◀

R Example

We now consider a classical example of seeing whether gender is independent of political affiliation in the USA. The data has numbers of males and females who feel they identify with a particular political affiliation and was considered in Agresti, A. (2007). An Introduction to Categorical Data Analysis, 2nd ed. New York: John Wiley & Sons. The data can be imputed directly into R as follows:

```
M <- as.table(rbind(c(762, 327, 468), c(484, 239, 477)))
dimnames(M) <- list(gender = c("F", "M"),
                party = c("Democrat","Independent",
                              "Republican"))
```

and looks like

	Democrat	Independent	Republican
F	762.00	327.00	468.00
M	484.00	239.00	477.00

To perform a chi-squared test to test the hypothesis that gender is independent of political affiliation, we use

```
chisq.test(M)
```

and we obtain the following output:

```
    Pearson's Chi-squared test

data:   M
X-squared = 30.07, df = 2, p-value = 2.954e-07
```

Here, the value of the test statistic is reported, the degrees of freedom (df) and the p-value of the test. It can be seen that there is strong evidence to reject the null hypothesis of independence between gender and political affiliation.
◀

5.4   Exercises

1. Two manufacturers of quartz watches (Romex and Tiddot) were tested for durability. The number of days to failure for a sample of size 5 of both makes is given below:

$$\text{Romex: } 101.4, 91.3, 108.0, 104.8, 102.2$$
$$\text{Tiddot: } 118.7, 102.5, 99.4, 129.1, 110.1$$

a. Compute the sample mean and sample variance of Romex and Tiddot watches from the samples given.
b. Let μ_X denote the population mean durability of Romex watches, and let μ_Y denote the population mean durability of Tiddot watches. Test the hypothesis that $\mu_X = \mu_Y$ against the alternative that $\mu_X \neq \mu_Y$ with a 5% significance level. You may assume that both samples come from populations which have equal variances. State any assumptions made and give the approximate p-value of the test.

2. The exam marks of a random sample of students studying economics were found to be as follows: 20, 34, 79, 94, 41, 42, 50. The economics department wishes to know if the means of all exam results were significantly higher than the pass mark of 40. Perform a suitable hypothesis test at a 5% significance level.
Following the introduction of a student support service, a second random sample of seven different students was taken, with the following results: 19, 20, 34, 94, 99, 81, 41. At a 5% significance level test the hypothesis

$$H_0 : \mu_X = \mu_Y$$
$$H_1 : \mu_X \neq \mu_Y$$

where μ_X and μ_Y are the mean results for the students without and with the student support service, respectively, assuming that the two populations of students have equal variances. State any other assumptions you have made.

3. The student support service claims in a student newspaper that 60% of students that use the service find it benefits their studies. An independent survey of 200 students who have used the student support service found that 132 out of the 200 stated that use of the service did improve their studies. Using the results of this independent survey, test the validity of the claim that 60% of students that use the student support service think it benefits their studies at a 5% significance level.

4. The numbers of a certain kind of crustacean were counted at 20 one-square-meter sites along a coastline, 10 facing South and 10 facing North. The results were as follows:

South:	28, 27, 31, 45, 21, 10, 30, 32, 25, 38
North:	27, 15, 38, 16, 21, 18, 20, 26, 30, 21

Using a 5% level of significance:

a. Test whether the population variances at each site are equal.
b. Test whether the population means at each site are equal, using the result of the previous question to guide your answer.

5. A researcher for a harbour control commission wishes to decide whether or not the number of ships that arrive in the harbour follows a Poisson distribution with mean $\mu = 1.0$. The researcher counted the number of ships arriving each hour for a random sample of 70 h and obtained the sample frequencies as given in the below table:

No. of ships arriving	Frequency
0	42
1	21
2	7
3 or more	0

Test the hypothesis that the number of ships that arrive in the harbour follows a Poisson distribution with mean $\mu = 1.0$ at a 5% significance level. You may find it helpful to recall that the probability mass function of a Poisson distribution with mean μ is given by

$$f(x; \mu) = \frac{\exp(-\mu)\mu^x}{x!}, \quad x = 0, 1, \ldots.$$

6. Let $X_1, X_2, \ldots, X_n$ be a random sample from a Normal distribution with mean μ and variance 1. We wish to test the hypothesis $H_0 : \mu = 0$ against $H_1 : \mu \neq 0$ at the 5% level of significance. Sketch the power function for sample sizes of 25, 100, and 500.

7. Suppose the random variable X is uniformly distributed on $[0, \theta]$. Let Z denote the maximum of a random sample of five observations from X. We wish to test the hypotheses $H_0 : \theta = 1$ against the alternative $H_1 : \theta < 1$ using the critical region $C = \{z : z \leq 0.5\}$. Find the power function for this hypothesis test and compute its significance level.

One-Way Analysis of Variance (ANOVA)

6

6.1 Introduction

In this chapter, a method for the analysis of an experiment that has more than two groups of observations is described. The objective is to determine if there are significant differences among the population means of the groups, which are assumed to be independent random samples from Normally distributed populations. The analysis is based on an examination of variation between and within groups, and is often called the one-way analysis of variance (or one-way ANOVA for short). What follows can be viewed as an extension of hypothesis tests for two means, to an experiment where there are more than two groups to be compared.

A valid question is "why do we need to construct another hypothesis test to compare the difference of three or more means, when we can do lots of tests comparing pairs of means?". Remember that every single statistical test that you perform runs the risk of making a Type I error. If you do lots of hypothesis tests on the same data, then you accumulate more and more risk of making a Type I error. Hence, if we can do one statistical test that answers all our hypotheses simultaneously, this is often preferable.

6.2 Notation and Setup

We introduce the following notation:

- There are m groups of data.
- There are n_i observations in the ith group.
- The jth observation in the ith group is denoted $x_{ij}, i = 1, \ldots, m, j = 1, \ldots, n_i$.

© Springer Nature Switzerland AG 2020
J. Gillard, *A First Course in Statistical Inference*,
Springer Undergraduate Mathematics Series,
https://doi.org/10.1007/978-3-030-39561-2_6

The data might be arranged as follows:

Group 1:	x_{11}	x_{12}	$\cdots$	x_{1n_1}
Group 2:	x_{21}	x_{22}	$\cdots$	x_{2n_2}
$\vdots$	$\vdots$	$\vdots$	$\vdots$	$\vdots$
Group m:	x_{m1}	x_{m2}	$\cdots$	x_{mn_m}

We assume that the above data is a random sample, where each x_{ij} is an observed value of a random variable X_{ij}. We assume that all the random variables X_{ij} are independent and have the Normal distribution with mean μ_i and common variance σ^2. The null hypothesis is $H_0 : \mu_1 = \mu_2 = \ldots = \mu_m$, where μ_i is the population mean corresponding to the ith group. The alternative hypothesis is $H_1 : \mu_a \neq \mu_b$, where $a \neq b$.

Let us introduce the following quantities, which will be the statistics used to test our hypotheses. We denote the total number of observations as $N = \sum_{i=1}^{m} n_i$.

The sum of the observations in the ith group is $x_{i\bullet} = \sum_{j=1}^{n_i} x_{ij}$, and the mean of the observations in the ith group is $\bar{x}_{i\bullet} = \dfrac{x_{i\bullet}}{n_i}$. The sum of all of the observations is given by $x_{\bullet\bullet} = \sum_{i=1}^{m} \sum_{j=1}^{n_i} x_{ij}$ and so we can write the overall mean of all the observations as $\bar{x}_{\bullet\bullet} = \dfrac{x_{\bullet\bullet}}{N}$.

The above notation may look strange. The bold dot $\bullet$ indicates that we are summing over the index where it is placed (the first index refers to the group number, and the second index refers to the observation within the group referenced by the first index).

6.3   Possible Sources of Variation

Let us try and visualize what our data may look like. Figure 6.1, for $m = 3$, helps us to identify all the sources of variation possible for our data. We can see three sources of variation, which are as follows:

1. Total variation (long dashed line), the overall amount of variability in the data;
2. Between-group variation (dotted line): how much do the groups vary from each other?
3. Within-group variation (solid line): how much does the data vary within each group?

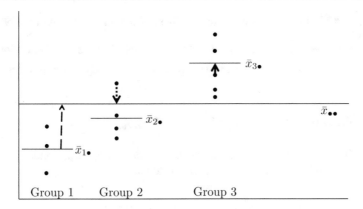

Fig. 6.1 Sources of variation in an ANOVA with $m = 3$ groups

Suppose we wish to quantify and compare the variation from each of these sources. We do this by building statistics to allow us to do this. We begin by introducing a "sum of squares" for each of these sources which are as follows:

1. Total sum of squares;
2. Between-group sum of squares;
3. Within-group sum of squares (also known as error, or residual sum of squares).

6.3.1 Total Sum of Squares, SS_{total}

To quantify the overall amount of variability in the data, we introduce a sum of squares denoted by SS_{total} which can be written as

$$SS_{total} = \sum_{i=1}^{m} \sum_{j=1}^{n_i} (x_{ij} - \bar{x}_{\bullet\bullet})^2 = \sum_{i=1}^{m} \sum_{j=1}^{n_i} x_{ij}^2 - \frac{x_{\bullet\bullet}^2}{N}.$$

Since each X_{ij} is Normally distributed with common variance σ^2, from Lemma 2.2 the random variable $\dfrac{\sum_{i=1}^{m} \sum_{j=1}^{n_i} (X_{ij} - \bar{X}_{\bullet\bullet})}{\sigma^2}$ can be seen to follow the chi-squared distribution with $N - 1$ degrees of freedom. Hence, we say that SS_{total} has $df_{total} = N - 1$ degrees of freedom (df).

6.3.2 Between-Group Sum of Squares, SS_G

To quantify the amount of variability between each group, we introduce a sum of squares denoted by SS_G which can be written as

$$SS_G = \sum_{i=1}^{m} \sum_{j=1}^{n_i} (\bar{x}_{i\bullet} - \bar{x}_{\bullet\bullet})^2 = \sum_{i=1}^{m} \frac{x_{i\bullet}^2}{n_i} - \frac{x_{\bullet\bullet}^2}{N}.$$

Following a similar argument to above, from Lemma 2.2 the random variable $\frac{\sum_{i=1}^{m}\sum_{j=1}^{n_i}(\bar{X}_{i\bullet} - \bar{X}_{\bullet\bullet})}{\sigma^2}$ can be seen to follow the chi-squared distribution with $m - 1$ degrees of freedom. Hence, we say that SS_G has $df_G = m - 1$ degrees of freedom.

6.3.3   Within-Group Sum of Squares, SS_{error}

To quantify the amount of variability within each group, we introduce a sum of squares denoted by SS_{error} which can be written as

$$SS_{error} = \sum_{i=1}^{m}\sum_{j=1}^{n_i}(x_{ij} - \bar{x}_{i\bullet})^2 = \sum_{i=1}^{m}\sum_{j=1}^{n_i}x_{ij}^2 - \sum_{i=1}^{m}\frac{x_{i\bullet}^2}{n_i}.$$

From Lemma 2.2, the random variable $\frac{\sum_{j=1}^{n_i}(X_{ij} - \bar{X}_{i\bullet})^2}{\sigma^2}$ can be seen to follow the chi-squared distribution with $n_i - 1$ degrees of freedom. Hence, the random variable $\frac{\sum_{i=1}^{m}\sum_{j=1}^{n_i}(X_{ij} - \bar{X}_{i\bullet})}{\sigma^2}$ follows the chi-squared distribution with $\sum_{i=1}^{m}n_i - 1 = N - m$ degrees of freedom. We say SS_{error} has $df_{error} = N - m$ degrees of freedom.

6.3.3.1 Relations Between Sums of Squares and Their Degrees of Freedom

It can be seen from inspection that $df_{total} = df_G + df_{error}$. This is an important computational shortcut: one needs to compute two degrees of freedom and then can infer the remaining one. This also translates to the sums of squares we have defined, and this is proven below.

Theorem 6.1 *Consider SS_{total}, SS_G, and SS_{error} as defined above. Then*

$$SS_{total} = SS_G + SS_{error}.$$

Proof

$$SS_{total} = \sum_{i=1}^{m}\sum_{j=1}^{n_i}(x_{ij} - \bar{x}_{\bullet\bullet})^2$$

$$= \sum_{i=1}^{m}\sum_{j=1}^{n_i}(x_{ij} - \bar{x}_{i\bullet} + \bar{x}_{i\bullet} - \bar{x}_{\bullet\bullet})^2$$

$$= \sum_{i=1}^{m}\sum_{j=1}^{n_i}(x_{ij} - \bar{x}_{i\bullet})^2 + \sum_{i=1}^{m}\sum_{j=1}^{n_i}(\bar{x}_{i\bullet} - \bar{x}_{\bullet\bullet})^2 + 2\sum_{i=1}^{m}\sum_{j=1}^{n_i}(x_{ij} - \bar{x}_{i\bullet})(\bar{x}_{i\bullet} - \bar{x}_{\bullet\bullet})$$

$$= SS_{error} + SS_G + 2\sum_{i=1}^{m}\sum_{j=1}^{n_i}(x_{ij} - \bar{x}_{i\bullet})(\bar{x}_{i\bullet} - \bar{x}_{\bullet\bullet})$$

Consider the final term. We can write

$$\sum_{j=1}^{n_i}(x_{ij}-\bar{x}_{i\bullet})(\bar{x}_{i\bullet}-\bar{x}_{\bullet\bullet}) = \sum_{j=1}^{n_i}x_{ij}(\bar{x}_{i\bullet}-\bar{x}_{\bullet\bullet}) - \sum_{j=1}^{n_i}\bar{x}_{i\bullet}(\bar{x}_{i\bullet}-\bar{x}_{\bullet\bullet})$$
$$= n_i\bar{x}_{i\bullet}(\bar{x}_{i\bullet}-\bar{x}_{\bullet\bullet}) - n_i\bar{x}_{i\bullet}(\bar{x}_{i\bullet}-\bar{x}_{\bullet\bullet}) = 0.$$

Hence $SS_{total} = SS_G + SS_{error}$. □

6.3.4   ANOVA Table

The comparison of sums of squares is often done in a so-called analysis of variance (ANOVA) table which is a succinct way of summarizing all sources of variation.

Source of variation	df	SS	MS	F-ratio
Between groups	$df_G = m - 1$	SS_G	$s_G^2 = \dfrac{SS_G}{df_G}$	$\dfrac{s_G^2}{s^2}$
Within groups (error or residual)	$df_{error} = N - m$	SS_{error}	$s^2 = \dfrac{SS_{error}}{df_{error}}$	–
Total	$df_{total} = N - 1$	SS_{total}	–	–

The mean squares (MS) are calculated by dividing the sum of squares (SS) by degrees of freedom (df). This isn't something we will discuss in depth, but each of the mean squares given above is unbiased estimator of the population variance for that source of variation. For example, s_G^2 is an unbiased estimator of the population between-groups variation. The F-ratio is the ratio of between-groups mean square to error mean square.

The idea is that if there is larger variance between sample means and small variance within samples, then the hypothesis of equal population means is less likely. The F-ratio is precisely the ratio of the statistics representing these two sources of variation.

Lemma 6.1 *Under the null hypothesis $H_0 : \mu_1 = \mu_2 = \ldots = \mu_m$, then the F-ratio $\dfrac{s_G^2}{s^2}$ follows an F-distribution with $(m-1)$ and $(N-m)$ degrees of freedom.*

The proof of the lemma follows from recognizing, as discussed earlier when the three sources of variation were described, that each of the statistics s_G^2 and s^2 follows chi-squared distributions, and so by Lemma 2.3 their ratio follows a F-distribution with degrees of freedom as given.

The percentiles of the F-distribution with $(m - 1)$ and $(N - m)$ degrees of freedom are used to judge the significance of the observed F-ratio. If the observed F-ratio exceeds the tabulated 95th percentile, for example, then with a 5% level of significance the null hypothesis (of equal population means) is rejected. More formally, the acceptance region $\mathcal{A}$ with a significance level α is

$$\mathcal{A} = \{f : 0 \le f \le F_{1-\alpha}(m - 1, N - m)\}$$

where the test statistic f is given by $f = \dfrac{s_G^2}{s^2}$ as described in the ANOVA table.

6.3.5 Multiple Comparison Tests

Although a one-way ANOVA can be used to identify that there are statistically significant differences in the population means of m groups of data, it does not identify which particular groups differ in this respect from the others. Various tests have been suggested to do this. One of the simplest to describe is Fisher's least significant difference (LSD) test.

Assume that each group of data is a random sample from an independent Normally distributed population. Then since $\bar{X}_{i\bullet}$ is a linear combination of Normally distributed variables, it is itself Normally distributed. For the kth group and lth group, one can show that

$$\frac{\bar{X}_{k\bullet} - \bar{X}_{l\bullet}}{\sqrt{s^2 \left(\frac{1}{n_k} + \frac{1}{n_l}\right)}} \sim t(N - m)$$

with $s^2 = \dfrac{SS_{error}}{N - m}$ as given in the ANOVA table.

Hence, the kth and lth groups will be deemed to have significantly different population means with significance level α if

$$|\bar{x}_{k\bullet} - \bar{x}_{l\bullet}| > t_{1-\alpha/2}(N - m)\sqrt{s^2 \left(\frac{1}{n_k} + \frac{1}{n_l}\right)}.$$

So if an ANOVA says that there are significant differences among the population means of the groups, we can then use Fisher's least significant difference test to see which population means differ from which.

Example

Systolic blood pressure was measured for seven human subjects, with five replicate observations being made for each subject. The results obtained (in mmHg) are tabulated below, together with some totals.

Subject	Systolic blood pressure	Totals
1	108 104 108 120 108	548
2	118 104 118 120 128	588
3	124 118 120 122 124	608
4	126 122 120 120 132	620
5	128 124 118 112 142	624
6	130 128 138 116 124	636
7	152 128 132 150 142	704
		4328

Complete an analysis of variance table for these data and test if there are significant differences among the population means of the subjects. Follow up your analysis by attempting to identify which groups differ from which. Use a 5% significance level.

Solution: We begin by computing sums of squares. We only need to compute two sums of squares; we can then work out what the remaining sum of squares is.

To begin

$$SS_{total} = \sum_{i=1}^{m} \sum_{j=1}^{n_i} x_{ij}^2 - \frac{x_{\bullet\bullet}^2}{N} = 539728 - \frac{4328^2}{35} = 4539.8857$$

which has $df_{total} = N - 1 = 35 - 1 = 34$ degrees of freedom.

Next

$$SS_G = \sum_{i=1}^{m} \frac{x_{i\bullet}^2}{n_i} - \frac{x_{\bullet\bullet}^2}{N} = \frac{548^2}{5} + \frac{588^2}{5} + \cdots + \frac{704^2}{5} - \frac{4328^2}{35} = 2731.8857$$

which has $df_G = m - 1 = 7 - 1 = 6$ degrees of freedom.

We can thus complete the ANOVA table as follows:

Source of variation	df	SS	MS	F-ratio
Between groups	6	2731.8857	455.3143	7.05
Within groups	28	1808.0000	64.5714	–
Total	34	4539.8857	–	–

For a 5% significance level, the acceptance region is given by

$$\mathcal{A} = \{f : 0 \le f \le F_{1-\alpha}(m - 1, N - m)\} = \{f : 0 \le f \le F_{0.95}(6, 28)\}$$
$$= \{f : 0 \le f \le 2.445\} .$$

Our test statistic $f = 7.05$ does not lie in this acceptance region and so we reject the null hypothesis of equal population means at a 5% significance level.

To detect where the differences lie, we will use Fisher's least significant difference test. The sample means (in ascending order) for each of the subjects are given by

$\bar{x}_{1\bullet}$	$\bar{x}_{2\bullet}$	$\bar{x}_{3\bullet}$	$\bar{x}_{4\bullet}$	$\bar{x}_{5\bullet}$	$\bar{x}_{6\bullet}$	$\bar{x}_{7\bullet}$
109.6	117.6	121.6	124.0	124.8	127.2	140.8

Two groups will be deemed to have significantly different population means at a 5% significance level if the absolute value of the difference between the sample means exceeds

$$t_{1-\alpha/2}(N-m)\sqrt{s^2\left(\frac{1}{n_k}+\frac{1}{n_l}\right)} = t_{0.975}(28)\sqrt{64.5714\left(\frac{1}{5}+\frac{1}{5}\right)} = 10.4083,$$

where $t_{0.975}(28) = 2.048$. For this example, all of the group sizes are equal, and so the problem is reduced to comparing all pairwise differences of the sample means with the above quantity.

Subjects 2 to 6 can be grouped as not having significantly different population means as $127.2 - 117.6 = 9.6 < 10.4083$. Subject 7 has blood pressure significantly higher than all other subjects. Subject 1 has a blood pressure significantly lower than all other subjects, apart from subject 2.

◀

R Example

We will consider an example to illustrate how to perform an ANOVA in R. We will consider the `mtcars` data set which can be loaded into R as follows:

```
data(mtcars)
attach(mtcars)
```

This data (which comes inbuilt in R) is from the 1974 Motor Trend US magazine and comprises fuel consumption (miles per gallon, mpg) and ten other variables for 32 cars. We can see the first six rows of the data set by using the following command:

```
head(mtcars, 6)
```

which outputs the following:

	mpg	cyl	disp	hp	drat	wt	qsec	vs	am	gear	carb
Mazda RX4	21.00	6.00	160.00	110.00	3.90	2.62	16.46	0.00	1.00	4.00	4.00
Mazda RX4 Wag	21.00	6.00	160.00	110.00	3.90	2.88	17.02	0.00	1.00	4.00	4.00
Datsun 710	22.80	4.00	108.00	93.00	3.85	2.32	18.61	1.00	1.00	4.00	1.00
Hornet 4 Drive	21.40	6.00	258.00	110.00	3.08	3.21	19.44	1.00	0.00	3.00	1.00
Hornet Sportabout	18.70	8.00	360.00	175.00	3.15	3.44	17.02	0.00	0.00	3.00	2.00
Valiant	18.10	6.00	225.00	105.00	2.76	3.46	20.22	1.00	0.00	3.00	1.00

Let us suppose we are interested how the average mpg depends on the number of gears. The data shows that cars either have 3, 4, or 5 gears. A boxplot of mpg by number of gears would be a good visualization to start. The code

```
boxplot(mpg ~ as.factor(gear), main = "mpg by No. of gears",
                    xlab = "Gears", ylab = "mpg")
```

produces the figure below:

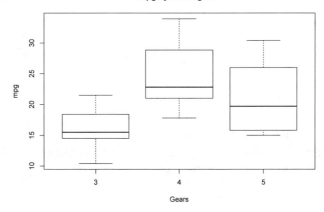

Suppose we wish to test the hypothesis that the average mpgs by number of gears are the same, against the alternative that they are not. We can test this hypothesis by using an ANOVA. We perform an ANOVA on mpg in R, by treating the number of gears (3, 4, or 5) as our different groupings using

```
cars<-aov(mpg ~ as.factor(gear))
summary(cars)
```

which outputs the following:

	Df	Sum Sq	Mean Sq	F value	Pr(>F)
as.factor(gear)	2	483.24	241.62	10.90	0.0003
Residuals	29	642.80	22.17		

In the final column, the p-value is reported, and so there can be seen to be much strong evidence to reject the null hypothesis of equal means. We can perform Fisher's LSD test to detect where the differences lie.

To perform Fisher's LSD test, we need to load the agricolae package into R. If it hasn't already been installed, then do so using

```
install.packages("agricolae")
```

first. Otherwise run

```
library("agricolae")
```

and then we can perform Fisher's LSD test on the already run ANOVA as follows:

```
out<-LSD.test(cars, "as.factor(gear)")
```

and using the code

```
print(out$groups)
```

we obtain

	mpg	groups
4	24.53	a
5	21.38	a
3	16.11	b

This table summarized the results of applying Fisher's LSD test at a 5% significance level: there are no significant differences between cars with 4 or 5 gears, but there is a significant difference of cars with 3 gears and those with 4 or 5 gears.

6.4   Exercises

1. The data below are the yields of four varieties of oats, each planted in three fields. Complete an analysis of variance table for these data and test to determine if there are significant differences among the means of the varieties. Follow up your analysis by using an appropriate method to identify which groups differ from which (using a 5% level of significance).

Variety	Yield		
1	10	20	0
2	200	0	100
3	90	800	700
4	100	20	40

2. An investigation was made of the concentration of a particular species of bacterium sometimes found in freshwater. Samples were taken from four different locations, and the concentrations (in millions of bacteria per liter of water) are tabulated below:

Location 1:	4.14	4.15	4.12
Location 2:	4.91	5.00	5.07
Location 3:	4.27	4.27	4.43
Location 4:	4.09	4.23	4.12

Test to determine if there are significant differences among the mean concentrations from the four locations. Use a multiple comparison test to determine which, if any, of the sites have significantly higher concentrations. Use a 5% significance level throughout.

3. Derive each of the following identities which were stated earlier:

a. $SS_{total} = \sum_{i=1}^{m} \sum_{j=1}^{n_i} (x_{ij} - \bar{x}_{\bullet\bullet})^2 = \sum_{i=1}^{m} \sum_{j=1}^{n_i} x_{ij}^2 - \dfrac{x_{\bullet\bullet}^2}{N}$

b. $SS_G = \sum_{i=1}^{m} \sum_{j=1}^{n_i} (\bar{x}_{i\bullet} - \bar{x}_{\bullet\bullet})^2 = \sum_{i=1}^{m} \dfrac{x_{i\bullet}^2}{n_i} - \dfrac{x_{\bullet\bullet}^2}{N}$

c. $SS_{error} = \sum_{i=1}^{m} \sum_{j=1}^{n_i} (x_{ij} - \bar{x}_{i\bullet})^2 = \sum_{i=1}^{m} \sum_{j=1}^{n_i} x_{ij}^2 - \sum_{i=1}^{m} \dfrac{x_{i\bullet}^2}{n_i}$

Regression: Fitting a Straight Line

7

7.1 Introduction

We now consider data consisting of pairs of observations (x_1, y_1), (x_2, y_2), ..., (x_n, y_n) where the x and y values are related via a straight-line model of the form

$$Y_i = \alpha + \beta x_i + \varepsilon_i, \quad i = 1, \ldots, n,$$

where α (the intercept) and β (the slope) are unknown population parameters to be estimated and the ε_i's are independent and identically distributed random variables having zero mean and constant variance σ^2. We assume that our x measurements are recorded without error, but that our Y measurements are subject to random error, hence, the use of the capital letter Y in the above model. Y is a random variable because of the random variable ε. In general, the variable x is called the independent variable and the variable Y is called the dependent variable.

Examples of independent and dependent variables include the following:

x (independent variable)	Y (dependent variable)
Amount spent advertising a product	Sales of the product
Age	Height
Distance	Time to run distance
Weight of fertilizer	Tomato yield

The primary aim of regression is to estimate the parameters α and β. We consider one possible way of how to do this in the next section.

© Springer Nature Switzerland AG 2020
J. Gillard, *A First Course in Statistical Inference*,
Springer Undergraduate Mathematics Series,
https://doi.org/10.1007/978-3-030-39561-2_7

Fig. 7.1 A plot of typical
paired data suitable for the
fitting of a straight line

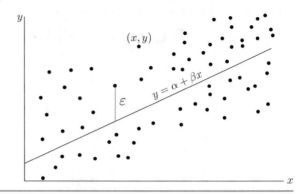

7.2   Least Squares Regression

A plot of typical data as described in the introduction is given in Fig. 7.1. We need
to fix some method to allow us to estimate α and β. Intuitively, we wish to find the
straight line which is "close" to as many of the observations $(x_1, y_1), (x_2, y_2), \ldots,$
(x_n, y_n) as possible. The sole source of error in our model is represented by the ε_i
terms; we assume that there are only errors associated with our dependent variable
Y. These errors are thus only in the vertical direction as illustrated.

We can thus begin by attempting to fit a straight line so that the vertical distance
between our observations and a line is as small as possible. One such method which
aims to do this is known as least squares.

The principle of least squares for our regression problem is to find α and β such
that the sum of squares

$$SS(\alpha, \beta) = \sum_{i=1}^{n}(y_i - \alpha - \beta x_i)^2$$

is as small as possible. The values of α and β which minimize the sum of squares
$SS(\alpha, \beta)$ are known as least squares estimates. We can find explicit formulae for
these least squares estimates, and this is the subject of the next theorem.

Theorem 7.1 *Suppose we have n pairs of observations* $(x_1, y_1), (x_2, y_2), \ldots,$
(x_n, y_n) *from the model*

$$Y_i = \alpha + \beta x_i + \varepsilon_i$$

*where α (the intercept) and β (the slope) are unknown constants and each ε_i is
a random error term with zero mean and constant variance σ^2. The least squares
estimators of α and β, denoted $\hat{\alpha}$ and $\hat{\beta}$, are given by*

$$\hat{\alpha} = \bar{y} - \hat{\beta}\bar{x},$$
$$\hat{\beta} = \frac{\sum_{i=1}^{n}(x_i - \bar{x})(y_i - \bar{y})}{\sum_{i=1}^{n}(x_i - \bar{x})^2}.$$

Proof Consider

$$SS(\alpha, \beta) = \sum_{i=1}^{n} (y_i - \alpha - \beta x_i)^2.$$

We wish to find estimates of α and β so that this sum of squares is as small as possible. Since $SS(\alpha, \beta)$ is a sum of squared terms, then $SS(\alpha, \beta) \geq 0$.

Let $u_i = x_i - \bar{x}$ and $v_i = y_i - \bar{y}$. Then $\sum_{i=1}^{n} u_i = \sum_{i=1}^{n} (x_i - \bar{x}) = \sum_{i=1}^{n} x_i - n\bar{x} = n\bar{x} - n\bar{x} = 0$. Similarly $\sum_{i=1}^{n} v_i = 0$. We will use these facts later in the proof.

We can write the following:

$$\sum_{i=1}^{n} (y_i - \alpha - \beta x_i)^2 = \sum_{i=1}^{n} (v_i + \bar{y} - \alpha - \beta(u_i + \bar{x}))^2$$

$$= \sum_{i=1}^{n} (v_i - \beta u_i - (\alpha + \beta\bar{x} - \bar{y}))^2$$

$$= \sum_{i=1}^{n} (v_i - \beta u_i)^2 - 2\sum_{i=1}^{n} (v_i - \beta u_i)(\alpha + \beta\bar{x} - \bar{y}) + \sum_{i=1}^{n} (\alpha + \beta\bar{x} - \bar{y})^2$$

$$= \sum_{i=1}^{n} (v_i - \beta u_i)^2 + 0 + n(\alpha + \beta\bar{x} - \bar{y})^2$$

The cross-product term is zero because $\sum_{i=1}^{n} u_i \sum_{i=1}^{n} v_i = 0$. This cross-product term, when expanded, can be written as $\sum_{i=1}^{n} u_i$ multiplied by something, plus $\sum_{i=1}^{n} v_i$ multiplied by something. Thus, our sum of squares $\sum_{i=1}^{n} (y_i - \alpha - \beta x_i)^2$ can be written as the sum of two terms, neither of which can be negative (because they are sums of squares themselves). The final term is minimized when it is zero. This is obtained when $\alpha = \bar{y} - \beta\bar{x}$. This is an estimate of the intercept and so we formally write $\hat{\alpha} = \bar{y} - \beta\bar{x}$.

It remains to obtain an estimator for β. Consider the first term $\sum_{i=1}^{n} (v_i - \beta u_i)^2$.

We can write this as follows:

$$\sum_{i=1}^{n} (v_i - \beta u_i)^2 = \sum_{i=1}^{n} v_i^2 - 2\beta \sum_{i=1}^{n} u_i v_i + \beta^2 \sum_{i=1}^{n} u_i^2$$

$$= \sum_{i=1}^{n} v_i^2 + \sum_{i=1}^{n} u_i^2 \left(\beta^2 - 2\beta \frac{\sum_{i=1}^{n} u_i v_i}{\sum_{i=1}^{n} u_i^2} \right)$$

$$= \sum_{i=1}^{n} v_i^2 + \sum_{i=1}^{n} u_i^2 \left[\left(\beta - \frac{\sum_{i=1}^{n} u_i v_i}{\sum_{i=1}^{n} u_i^2} \right)^2 - \left(\frac{\sum_{i=1}^{n} u_i v_i}{\sum_{i=1}^{n} u_i^2} \right)^2 \right]$$

This term is minimized when we choose $\beta = \hat{\beta} = \dfrac{\sum_{i=1}^{n} u_i v_i}{\sum_{i=1}^{n} u_i^2} = \dfrac{\sum_{i=1}^{n}(x_i - \bar{x})(y_i - \bar{y})}{\sum_{i=1}^{n}(x_i - \bar{x})^2}$　□

If we write

$$S_{xx} = \sum_{i=1}^{n}(x_i - \bar{x})^2 = \sum_{i=1}^{n}(x_i - \bar{x})x_i = \sum_{i=1}^{n} x_i^2 - \frac{\left(\sum_{i=1}^{n} x_i\right)^2}{n} = \sum_{i=1}^{n} x_i^2 - n\bar{x}^2$$

$$S_{xy} = \sum_{i=1}^{n}(x_i - \bar{x})(y_i - \bar{y}) = \sum_{i=1}^{n}(x_i - \bar{x})y_i = \sum_{i=1}^{n} x_i(y_i - \bar{y}) = \sum_{i=1}^{n} x_i y_i - \frac{\left(\sum_{i=1}^{n} x_i\right)\left(\sum_{i=1}^{n} y_i\right)}{n}$$

$$= \sum_{i=1}^{n} x_i y_i - n\bar{x}\bar{y}$$

where S_{yy} is defined analogously to S_{xx} with the obvious changes, then we can write our estimator of the slope $\hat{\beta}$ as

$$\hat{\beta} = \frac{\sum_{i=1}^{n}(x_i - \bar{x})(y_i - \bar{y})}{\sum_{i=1}^{n}(x_i - \bar{x})^2} = \frac{S_{xy}}{S_{xx}} = \frac{\sum_{i=1}^{n} x_i y_i - n\bar{x}\bar{y}}{\sum_{i=1}^{n} x_i^2 - n\bar{x}^2}.$$

By the principle of least squares, the minimum value of

$$SS(\alpha, \beta) = \sum_{i=1}^{n}(y_i - \alpha - \beta x_i)^2$$

is given by

$$SS(\hat{\alpha}, \hat{\beta}) = \sum_{i=1}^{n}(y_i - \hat{\alpha} - \hat{\beta} x_i)^2 = SS_{error},$$

and this is known as the residual sum of squares. We denote this by SS_{error}. It is a measure of how good our straight-line fit is to our observed data.

Lemma 7.1 *Let $\hat{\alpha}$ and $\hat{\beta}$ be defined as in Theorem 7.1. The residual sum of squares, SS_{error}, may be written as*

$$SS_{error} = SS(\hat{\alpha}, \hat{\beta}) = \sum_{i=1}^{n}(y_i - \hat{\alpha} - \hat{\beta} x_i)^2 = S_{yy} - \frac{S_{xy}^2}{S_{xx}}.$$

Try to prove this yourself. The residual sum of squares is an important quantity which we will use later.

Example

The electrical resistance y of a piece of copper wire depends linearly upon its temperature x. The following measurements of the resistance (in ohms) at different temperatures (in degrees Celsius) were obtained:

x	2	10	15	19
y	1.324	1.368	1.391	1.415

Find the least squares estimates of the slope and intercept of the straight line to predict y with x. Use the least square estimates to predict the electrical resistance when the temperature is $12\,°C$ and compute the residual sum of squares SS_{error}.

Solution: For the data given, we can evaluate the following statistics: $\bar{x} = 11.5$, $\bar{y} = 1.3745$, $\sum_{i=1}^{4} x_i^2 = 690$, $\sum_{i=1}^{4} x_i y_i = 64.078$, and $\sum_{i=1}^{4} y_i^2 = 7.5615$. Hence, we can evaluate the following important quantities:

$$S_{xx} = \sum_{i=1}^{n} x_i^2 - n\bar{x}^2 = 690 - (4 \times 11.5^2) = 161,$$

$$S_{xy} = \sum_{i=1}^{n} x_i y_i - n\bar{x}\bar{y} = 64.078 - (4 \times 11.5 \times 1.3745) = 0.851,$$

$$S_{yy} = \sum_{i=1}^{n} y_i^2 - n\bar{y}^2 = 7.5615 - (4 \times 1.3745^2) = 0.004505.$$

The least squares estimate of the slope of the straight line is given by

$$\hat{\beta} = \frac{S_{xy}}{S_{xx}} = \frac{0.851}{161} = 0.0052857$$

and the least squares estimate of the intercept of the straight line is given by

$$\hat{\alpha} = \bar{y} - \hat{\beta}\bar{x} = 1.3745 - 0.0052857 \times 11.5 = 1.3137.$$

Our regression equation is thus

$$y = \hat{\alpha} + \hat{\beta}x = 1.3137 + 0.0052857x.$$

The prediction of the electrical resistance when the temperature is $x = 12$ is found to be

$$\hat{y} = 1.3137 + 0.0052857 \times 12 = 1.3771.$$

The residual sum of squares, SS_{error}, can be found as

$$SS_{error} = S_{yy} - \frac{S_{xy}^2}{S_{xx}} = 0.004505 - \frac{0.851^2}{161} \approx 0.000007.$$

◄

7.3   Properties of the Least Squares Estimators: Distributions

Let Y_i denote the random variable whose observed value is y_i. It is a random variable due to the addition of the random error term ε_i. To consider properties of the estimator $\hat{\beta}$ (which will enable us to consider properties of the estimator $\hat{\alpha}$ as well as the predicted value of y given a value of x), we write

$$\hat{\beta} = \sum_{i=1}^{n} \frac{(x_i - \bar{x})(Y_i - \bar{Y})}{S_{xx}}$$

to emphasize that Y is a random variable.

Lemma 7.2 *We may write*

$$\hat{\beta} = \sum_{i=1}^{n} \frac{(x_i - \bar{x})(Y_i - \bar{Y})}{S_{xx}} = \sum_{i=1}^{n} \frac{(x_i - \bar{x})Y_i}{S_{xx}}.$$

Proof Consider the numerator, and expand

$$\sum_{i=1}^{n} (x_i - \bar{x})(Y_i - \bar{Y}) = \sum_{i=1}^{n} (x_i - \bar{x})Y_i - \bar{Y} \sum_{i=1}^{n} (x_i - \bar{x}) = \sum_{i=1}^{n} (x_i - \bar{x})Y_i$$

since $\sum_{i=1}^{n} (x_i - \bar{x}) = 0$. Hence $\hat{\beta} = \sum_{i=1}^{n} \frac{(x_i - \bar{x})(Y_i - \bar{Y})}{S_{xx}} = \sum_{i=1}^{n} \frac{(x_i - \bar{x})Y_i}{S_{xx}}$.   □

We use this form of the estimator to derive properties of $\hat{\beta}$ (and other parameters). In order to construct confidence intervals and hypothesis tests for estimators of the parameters of the model, and for predictions of y using it, we need to make some distributional assumptions concerning the random error term ε_i of the model:

1. They are all mutually independent.
2. They have zero expectation.
3. They all have the same variance σ^2.
4. They are Normally distributed.

We can derive the following facts using the above assumptions:

$$E[Y_i] = E[\alpha + \beta x_i + \varepsilon_i] = E[\alpha + \beta x_i] + E[\varepsilon_i] = \alpha + \beta x_i,$$
$$Var[Y_i] = Var[\alpha + \beta x_i + \varepsilon_i] = Var[\alpha + \beta x_i] + Var[\varepsilon_i] = \sigma^2.$$

The only random variables in our model are the ε_i's, and these are assumed to be independent. It follows the Y_i's are themselves also independent.

7.3.1   Mean and Variance of $\hat{\beta}$

We have that

$$E[\hat{\beta}] = E\left[\sum_{i=1}^{n} \frac{(x_i - \bar{x})Y_i}{S_{xx}}\right] = \sum_{i=1}^{n} \frac{(x_i - \bar{x})E[Y_i]}{S_{xx}}$$

$$= \sum_{i=1}^{n} \frac{(x_i - \bar{x})(\alpha + \beta x_i)}{S_{xx}}$$

$$= \frac{\alpha}{S_{xx}} \sum_{i=1}^{n} (x_i - \bar{x}) + \frac{\beta}{S_{xx}} \sum_{i=1}^{n} x_i(x_i - \bar{x})$$

$$= 0 + \frac{\beta}{S_{xx}} S_{xx} = \beta.$$

So we can say that our estimator $\hat{\beta}$ is an unbiased estimator of the true slope β. Its variance is as follows:

$$Var[\hat{\beta}] = Var\left[\sum_{i=1}^{n} \frac{(x_i - \bar{x})Y_i}{S_{xx}}\right] = \sum_{i=1}^{n} \frac{(x_i - \bar{x})^2 Var[Y_i]}{S_{xx}^2} = \frac{S_{xx}\sigma^2}{S_{xx}^2} = \frac{\sigma^2}{S_{xx}}.$$

We now consider the remaining parameters. The step to derive the mean and variance of these parameters is largely similar to what has been considered for $\hat{\beta}$.

7.3.2   Mean and Variance of $\hat{\alpha}$

The previous derivation was enabled by writing the estimator as multiples of our random variables Y_i. We then used properties concerning the expectation and variance of multiples of random variables to derive the expectation and variance of $\hat{\beta}$. Our first step is to write $\hat{\alpha}$ as a combination of the random variables Y_i:

$$\hat{\alpha} = \bar{Y} - \hat{\beta}\bar{x} = \sum_{i=1}^{n} \frac{Y_i}{n} - \frac{S_{xy}}{S_{xx}}\bar{x} = \sum_{i=1}^{n} \frac{Y_i}{n} - \frac{\bar{x}}{S_{xx}} \sum_{i=1}^{n} (x_i - \bar{x})Y_i = \sum_{i=1}^{n} \left(\frac{1}{n} - \frac{\bar{x}(x_i - \bar{x})}{S_{xx}}\right) Y_i.$$

This is the most difficult part of the job complete. Remembering that the random variables are the Y_i, we can now compute the expectation and variance of $\hat{\alpha}$ using the standard rules of the expectation and variance of a combination of random variables.

$$E[\hat{\alpha}] = \sum_{i=1}^{n} \left(\frac{1}{n} - \frac{\bar{x}(x_i - \bar{x})}{S_{xx}}\right) E[Y_i] = \sum_{i=1}^{n} \left(\frac{1}{n} - \frac{\bar{x}(x_i - \bar{x})}{S_{xx}}\right) (\alpha + \beta x_i)$$

$$= \sum_{i=1}^{n} \left(\frac{\alpha}{n} - \frac{\alpha\bar{x}(x_i - \bar{x})}{S_{xx}} + \frac{\beta x_i}{n} - \frac{\beta\bar{x}(x_i - \bar{x})x_i}{S_{xx}}\right)$$

$$= \alpha - 0 + \beta\bar{x} - \beta\bar{x} = \alpha.$$

So we can say that our estimator $\hat{\alpha}$ is an unbiased estimator of the true slope α. Its variance is as follows:

$$Var[\hat{\alpha}] = \sum_{i=1}^{n} \left(\frac{1}{n} - \frac{\bar{x}(x_i - \bar{x})}{S_{xx}} \right)^2 Var[Y_i] = \sum_{i=1}^{n} \left(\frac{1}{n} - \frac{\bar{x}(x_i - \bar{x})}{S_{xx}} \right)^2 \sigma^2 = \sigma^2 \left(\frac{1}{n} + \frac{\bar{x}^2}{S_{xx}} \right).$$

7.3.3 Mean and Variance of $\hat{y}_0$

Suppose we are given a value of x, say x_0. We may be asked to estimate the response y_0 at this given value. For example, given someone's height, could you estimate what their weight would be? According to our model, given x_0, our estimated response y_0 would be given by $\hat{y}_0 = \hat{\alpha} + \hat{\beta} x_0$. This is just an estimator, and so we can again consider the mean and variance of this estimator.

Again the first step is to write our estimator $\hat{y}_0$ as a combination of random variables:

$$\hat{y}_0 = \hat{\alpha} + \hat{\beta} x_0 = \bar{Y} - \hat{\beta}\bar{x} + \hat{\beta} x_0 = \bar{Y} + \hat{\beta}(x_0 - \bar{x}) = \sum_{i=1}^{n} \frac{Y_i}{n} + \frac{S_{xy}}{S_{xx}}(x_0 - \bar{x})$$

$$= \sum_{i=1}^{n} \frac{Y_i}{n} + \sum_{i=1}^{n} \frac{(x_i - \bar{x})Y_i(x_0 - \bar{x})}{S_{xx}}$$

$$= \sum_{i=1}^{n} \left(\frac{1}{n} + \frac{(x_i - \bar{x})(x_0 - \bar{x})}{S_{xx}} \right) Y_i ,$$

and now one can show that

$$E[\hat{y}_0] = \alpha + \beta x_0 = y_0, \quad Var[\hat{y}_0] = \sigma^2 \left(\frac{1}{n} + \frac{(x_0 - \bar{x})^2}{S_{xx}} \right).$$

7.3.4 Distributional Results

An examination of the expressions for $\hat{\beta}$, $\hat{\alpha}$, and $\hat{y}_0$ shows that they are all expressible as linear combinations of $Y_1, Y_2, \ldots, Y_n$. It follows that if the ε_i's are Normally distributed, then so are $\hat{\beta}$, $\hat{\alpha}$, and $\hat{y}_0$. In this case, we can summarize the preceding results by stating that

$$\hat{\beta} \sim N \left[\beta, \frac{\sigma^2}{S_{xx}} \right],$$

$$\hat{\alpha} \sim N \left[\alpha, \sigma^2 \left(\frac{1}{n} + \frac{\bar{x}^2}{S_{xx}} \right) \right],$$

$$\hat{y}_0 = \hat{\alpha} + \hat{\beta} x_0 \sim N \left[y_0, \sigma^2 \left(\frac{1}{n} + \frac{(x_0 - \bar{x})^2}{S_{xx}} \right) \right].$$

These can be used to make confidence intervals and hypothesis tests about these parameters in the same way as considered in Chaps. 4 and 5.

Example

Derive a $100(1 - \alpha)\%$ confidence interval for the unknown slope β of the model $Y_i = \alpha + \beta x_i + \varepsilon_i$ upon having observed a sample of n pairs of observations $(x_1, y_1), (x_2, y_2), \ldots, (x_n, y_n)$. Assume that the errors ε_i are independent and identically distributed Normal random variables with zero mean and known variance σ^2.

Solution: We have already noted above that

$$\hat{\beta} \sim N\left[\beta, \frac{\sigma^2}{S_{xx}}\right],$$

or by standardizing we obtain

$$\frac{\hat{\beta} - \beta}{\sqrt{\frac{\sigma^2}{S_{xx}}}} \sim N[0, 1].$$

Analogous to the derivation of a confidence interval for an unknown mean derived in Sect. 4.2.1.1, we can write the following probability statement:

$$P\left[-z_{1-\alpha/2} \leq \frac{\hat{\beta} - \beta}{\sqrt{\frac{\sigma^2}{S_{xx}}}} \leq z_{1-\alpha/2}\right] = (1 - \alpha)$$

or rearranging terms within this probability statement we obtain

$$P\left[\hat{\beta} - z_{1-\alpha/2}\sqrt{\frac{\sigma^2}{S_{xx}}} \leq \beta \leq \hat{\beta} + z_{1-\alpha/2}\sqrt{\frac{\sigma^2}{S_{xx}}}\right] = (1 - \alpha)$$

giving a $100(1 - \alpha)\%$ confidence interval for β to be $\hat{\beta} \pm z_{1-\alpha/2}\sqrt{\frac{\sigma^2}{S_{xx}}}$.

Similar derivations yield $100(1 - \alpha)\%$ confidence intervals for α and y_0 to be

$$\hat{\alpha} \pm z_{1-\alpha/2}\sqrt{\sigma^2\left(\frac{1}{n} + \frac{\bar{x}^2}{S_{xx}}\right)} \quad \text{and} \quad \hat{y}_0 \pm z_{1-\alpha/2}\sqrt{\sigma^2\left(\frac{1}{n} + \frac{(x_0 - \bar{x})^2}{S_{xx}}\right)},$$

respectively.

◄

7.3.5   Estimating the Error Variance σ^2

The value of the error variance σ^2 is usually unknown and must be estimated from the data. Recall that in Lemma 7.1 we stated that the residual sum of squares, SS_{error} may be written as

$$SS_{error} = SS(\hat{\alpha}, \hat{\beta}) = \sum_{i=1}^{n}(y_i - \hat{\alpha} - \hat{\beta}x_i)^2 = S_{yy} - \frac{S_{xy}^2}{S_{xx}}.$$

The residual sum of squares is important because of the next lemma; it is the key ingredient to estimate the variance of each of the ε_i terms, denoted σ^2.

Lemma 7.3

$$s^2 = \frac{SS_{error}}{n-2}$$

is an unbiased estimator of σ^2.

It follows from earlier results that if the errors are Normally distributed then the random variables

$$\frac{\hat{\beta} - \beta}{\sqrt{\frac{\sigma^2}{S_{xx}}}}, \quad \frac{\hat{\alpha} - \alpha}{\sqrt{\sigma^2\left(\frac{1}{n} + \frac{\bar{x}^2}{S_{xx}}\right)}}, \quad \frac{\hat{y}_0 - y_0}{\sqrt{\sigma^2\left(\frac{1}{n} + \frac{(x_0-\bar{x})^2}{S_{xx}}\right)}}$$

have the distribution $N[0, 1]$. It can be shown that when σ^2 is estimated by

$$\hat{\sigma}^2 = s^2 = \frac{SS_{error}}{n-2}$$

then the random variables

$$\frac{\hat{\beta} - \beta}{\sqrt{\frac{s^2}{S_{xx}}}}, \quad \frac{\hat{\alpha} - \alpha}{\sqrt{s^2\left(\frac{1}{n} + \frac{\bar{x}^2}{S_{xx}}\right)}}, \quad \frac{\hat{y}_0 - y_0}{\sqrt{s^2\left(\frac{1}{n} + \frac{(x_0-\bar{x})^2}{S_{xx}}\right)}}$$

each have the Student's t-distribution with $(n-2)$ degrees of freedom. Again these can be used to make confidence intervals, or to perform hypothesis tests in the usual way.

Example

The following table gives measurements of two variables x and y that are known to be linearly related. The variable x is measured without error, but there is a measurement error associated with y that you can assume is Normally distributed with zero mean and unknown variance σ^2.

x:	5.0	7.5	10.0	12.5	15.0
y:	1.23	1.39	1.52	1.66	1.81

Find the least squares estimates of the slope and intercept of the straight line to predict y with x. Also calculate an unbiased estimate of the error variance σ^2. Find a 95% confidence interval for the intercept.

Solution: We first compute the following simple statistics, $\bar{x} = 10$, $\bar{y} = 1.522$, $\sum_{i=1}^{n} x_i^2 = 562.5$, $\sum_{i=1}^{n} y_i^2 = 11.7871$, $\sum_{i=1}^{n} x_i y_i = 79.675$, which enables us to compute the following key quantities: $S_{xx} = \sum_{i=1}^{n} x_i^2 - n\bar{x}^2 = 62.5$, $S_{xy} = \sum_{i=1}^{n} x_i y_i - n\bar{x}\bar{y} = 3.575$, and $S_{yy} = \sum_{i=1}^{n} y_i^2 - n\bar{y}^2 = 0.20468$.

We estimate the slope and intercept to be

$$\hat{\beta} = \frac{S_{xy}}{S_{xx}} = 0.0572, \quad \hat{\alpha} = \bar{y} - \hat{\beta}\bar{x} = 0.95.$$

An unbiased estimator of σ^2 is

$$s^2 = \frac{SS_{error}}{n-2} = \frac{S_{yy} - \frac{S_{xy}^2}{S_{xx}}}{n-2} = 0.000063.$$

As the errors are assumed to be Normally distributed (also need to assume that they have zero mean, constant variance, and are independent) then

$$\frac{\hat{\alpha} - \alpha}{\sqrt{s^2 \left(\frac{1}{n} + \frac{\bar{x}^2}{S_{xx}} \right)}} \sim t(n-2)$$

and a 95% confidence interval for α is given by

$$\hat{\alpha} \pm t_{0.975}(n-2)\sqrt{s^2 \left(\frac{1}{n} + \frac{\bar{x}^2}{S_{xx}} \right)} = 0.95 \pm 3.182\sqrt{0.000063 \left(\frac{1}{5} + \frac{100}{62.5} \right)}$$

$$= [0.91602, 0.98398].$$

◄

R Example

We will consider an example to illustrate how to perform a linear regression in R by considering the `cars` data set which can be loaded into R as follows:

```
data(cars)
attach(cars)
```

This data (which comes inbuilt in R) consists of 50 observations of speed and corresponding distances. To see what the data looks like, we submit the following:

```
head(cars)
```

which outputs

	speed	dist
1	4.00	2.00
2	4.00	10.00
3	7.00	4.00
4	7.00	22.00
5	8.00	16.00
6	9.00	10.00

We can produce a scatterplot of the above data by using

```
plot(speed, dist, xlab="Speed", ylab="Distance")
```

which produces

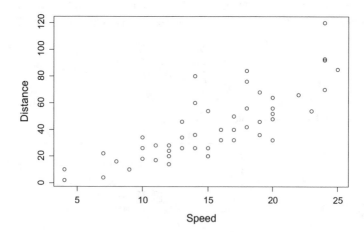

Let us suppose we are interested in fitting a straight line to the above data, with `speed` as the independent variable and `dist` as the independent variable. The code

```
regr <- lm(dist ~ speed)
summary(regr)
```

produces the following table:

	Estimate	Std. Error	t value	Pr(>|t|)
(Intercept)	−17.5791	6.7584	−2.60	0.0123
speed	3.9324	0.4155	9.46	0.0000

In summary, the model that has been fitted is $dist = -17.5791 + 3.9324 \times$ speed. The scatterplot with this straight line overlaid is given below:

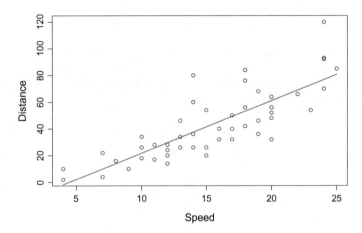

As well as the estimated parameters of the straight line, the R output gives the p-value of testing the hypothesis that each of the parameters is equal to zero. Inspecting the table shows that the p-values are very small, and there is much strong evidence to suggest that the intercept and the coefficient of speed (which is the slope of the estimated straight line) are significantly different from zero. We can obtain 95% confidence intervals for each of these parameters by using

```
confint(regr)
```

which produces the following output:

	2.5%	97.5%
(Intercept)	−31.17	−3.99
speed	3.10	4.77

which says that the 95% confidence intervals for the intercept and the slope of the line are $(-31.17, -3.99)$ and $(3.10, 4.77)$, respectively.

◀

7.4   Exercises

1. A firm has a fleet of six cars. It wishes to study the relationship between the fuel cost per mile (in pounds) of operating the car, against the car's age (in years). The data is given below:

Age (in years)	Fuel cost per mile (in pounds)
1	2.0
1	2.0
2	2.0
3	3.0
4	2.5
7	3.5

Fit a linear regression of fuel cost per mile against age. Use your regression to predict the expected fuel cost per mile for a car that is 2 years old.

2. The table below gives the lengths x and the head girths y of five squirrels, all measurements being in centimeters.

x	y
23	15.6
24	17.3
25	17.9
26	18.4
27	19.1

Calculate the least square estimates of the slope and intercept of a straight line to predict y with x, and test the null hypothesis that the slope of the line relating x and y is equal to zero at a 1% significance level. State any assumptions made.

3. The table below gives the lengths x and the head girths y of ten redfish, all measurements being in centimeters.

x	y
23	15.6
24	17.3
25	17.9
26	18.4
27	19.1
28	20.1
29	20.8
30	21.8
31	22.3
32	22.8

Calculate the least square estimates of the slope and intercept of a straight line to predict y with x, and test the null hypothesis that the slope of the line relating x and

y is equal to zero at a 1% significance level. Calculate the 95% confidence interval for the head girth of a fish of length 23 cm and comment on the measurement already made of a fish of length 23 cm.

A Brief Introduction to R

A

A.1   Introduction

This Appendix will get you started with R. Its aims is to support you to find the program, install it, and to begin experimenting with the software. The details here are not sufficient on their own. You will certainly have to find additional resources if you are new to programming. Luckily, there are a huge number of resources available online, as well as an abundance of books introducing you to R. Indeed, as your expertise with R grows you will see that there are whole textbooks dedicated to specific aspects of R and how to perform certain analyses or methods using R. This demonstrates the power of the software, and also its popularity.

R is open-source software freely available to anyone. Many declare it to be the standard software package for academic statisticians, and it is gaining ever-increasing support in industry. It now forms a key part of mathematics and statistics degree schemes and is also a common tool taught in sciences where the analysis of data is common, such as biology and psychology.

R is available for download (for Windows, MacOS, and UNIX) at https://www.r-project.org/, which also contains a number of get started tutorials describing how to install the software and how to begin performing computations with R. This is a good place to start. There are a number of additional interfaces which provide other means of interacting with R. One example is RStudio available at https://rstudio.com/.

A.2   What R Looks Like

Assuming that you have installed R correctly, upon opening the software you will see something like that presented in Fig. 1.

Note that depending on your operating system and computer, the precise look of R might be different from above. You can directly type commands into the console window and press enter to execute them. Try for example

© Springer Nature Switzerland AG 2020

J. Gillard, *A First Course in Statistical Inference*,
Springer Undergraduate Mathematics Series,
https://doi.org/10.1007/978-3-030-39561-2_A

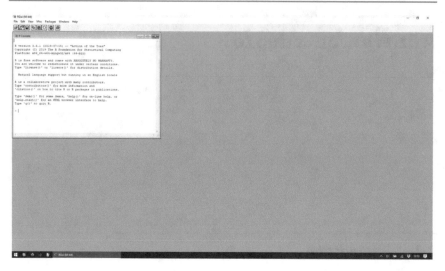

Fig. 1 Opening R

```
2+2
```

which when executed will give the output

```
[1]  4
```

We can assign a label to a variable if we wish. For example, we can assign the value "5.3" to a variable called x by using

```
x <-  5.3
```

and then to find x^2 and save it to a new variable y we can use

```
y <-  x^2
```

To print this variable we can execute

```
print(y)
[1]  28.09
```

Here are some other examples which you should try inputting and executing yourself:

Input	Meaning
`a <- 1:6`	Make a vector $(1, 2, 3, 4, 5, 6)$ and assign it to a
`mean(a)`	Compute the arithmetic mean of a
`var(a)`	Compute the sample variance of a
`names <- c("Bob","Billy")`	Make a vector of names (*Bob*, *Billy*) and assign it to names
`age <- c(31,23)`	Make a vector of ages $(31, 23)$ and assign it to age

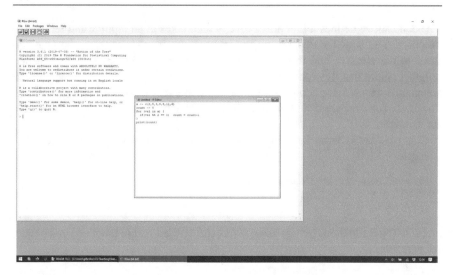

Fig. 2 A script file in R

Typing in commands and executing them in the console window is likely to not be suitable if we need to submit multiple commands, and it is not obvious how we may save what we have executed. To get around this we can create a script file and edit it in the editor window, highlighting specific text we wish to run. This approach has the benefit of being able to save the commands written in the script files and is the most common way R is used.

A script file looks as given in Fig. 2, and one is opened from the menu by following "File → New script". When writing scripts it is good practice to include comments in our files that help describe what the code does. The way to do this in R is with the # symbol before text. Any text which follows # is ignored by R and this is a useful way to save commentary explaining what your R code is attempting to do. This is both for other possible users of what you have written and yourself so that you understand code that you have written when you return to it (perhaps much) later.

Functions and Packages

There are two key features of R which make it popular among practitioners and academics:

1. The ability to write bespoke functions in R,
2. The option of installing additional packages which typically perform some advanced analysis or task using functions written by others.

To introduce functions, let us discuss some examples.

```
fun1 <- function(x){
    r <- if (x<0) x+3 else x+2
    return(r)
}
```

The code above defines a function `fun1`, which takes in an input x and returns an output r. The output r depends on the input. If the input is negative then we add three to the number, and then output it. Otherwise, we add two to the input and output it. Note that this function also introduces a conditional `if` statement. Such statements which place conditions on the input are very common in programming. To run the function we would execute `fun1(10)` which would give the output `12`, for example.

It is possible to write functions containing any number of input arguments, for example,

```
sum.of.squares <- function(x,y) {
  x^2 + y^2
}
```

and existing R functions (in this case `plot`) can be used within functions you create, for example,

```
red.plot <- function(x, y) {
  plot(x, y, col="red")
}
```

A useful tool in R is a `for` loop which is used to iterate a function over multiple inputs. Here is an example:

```
x <- c(2,5,3,9,8,11,6)
count <- 0
for (val in x) {
  if(val %% 2 == 0)  count = count+1
}
print(count)
```

Here, we define a vector x to be the numbers as given and set the initial value of a variable called `count` to be 0. We then use the `for` statement to take every value in x (which we denote by `val`) and check if it is even by seeing if the remainder of `val` upon division by 2 is 0. This is done by the modulo arithmetic `%%` command. Note that within an `if` statement, to assess if a variable is equal to some number we use a double equals `==` to avoid confusion with a single equal sign. If `val` is even then we update our variable `count` by adding 1 to it. Given that the starting value of `count` is 0, this program thus returns the number of even numbers in the vector x. We can generalize the above code to operate on any vector x by making our own function:

```
count.even.numbers <- function(x) {
   count <- 0
   for (val in x) {
     if(val %% 2 == 0)  count = count+1
   }
   print(count)
}
```

In R, the fundamental unit of shareable code is the so-called package. A package is a bundle containing code, data, documentation, and tests. There are over 6,000 packages available on the CRAN, which is the public repository for R packages. The availability and variety of packages is one of the reasons that R is so successful: the chances are that someone has already solved a problem that you're working on, and you can benefit from their work by downloading and using their package. Here is the important code for using packages in R:

- The package named "knitr" is installed by using the code `install.packages("knitr")`. Packages only need to be installed on your machine once.
- To load the package "knitr" into R we use `library("knitr")`.
- To get help on the package "knitr" we use `package?knitr` and `help(package="knitr")`.

A.3   Dealing with Data

Manipulating and performing analyses of data is a key functionality of R. There are ways of importing data from external formats (such as Excel or text files) into R, and there are a number of inbuilt data sets in R one can use to experiment with. One such inbuilt data set that is used in this book is the so-called iris data set, which gives measurements in centimeters of the sepal length, sepal width, petal length, and petal width for 50 flowers from three species of iris.

To load this data into R we run `data("iris")`. We can inspect the top section of data by running `head(iris)` which produces

	Sepal.Length	Sepal.Width	Petal.Length	Petal.Width	Species
1	5.10	3.50	1.40	0.20	setosa
2	4.90	3.00	1.40	0.20	setosa
3	4.70	3.20	1.30	0.20	setosa
4	4.60	3.10	1.50	0.20	setosa
5	5.00	3.60	1.40	0.20	setosa
6	5.40	3.90	1.70	0.40	setosa

We can then ask for the mean of all sepal lengths by using `mean(iris$Sepal.Length)` which gives the answer 5.843333. Here before the $ sign we give the data set we are studying, and after the $ sign the specific variable within that data set we would like to compute the mean of. This method of referring to variables within data sets is standard in R and allows for comparisons of variables across different data sets.

Suppose, however, that we wish to study just one data set. We can fix the data set by using the `attach(iris)` command. This means that R assumes that whenever we refer to a variable, it should be taken from the data set we have attached. So once we have attached the iris data set, we can run commands such as `mean(Sepal.Length)` which would give us the same answer as before.

Solutions to Exercises

B

B.1   Problems of Chap. 1

1. To find k we can use the property that the probabilities (as given by the probability mass function) must sum to 1. In this case

$$f(1) + f(2) + f(3) = k + 4k + 9k = 14k = 1,$$

and so $k = 1/14$. We can therefore write that the probability mass function is

$$f(x) = P[X = x] = \begin{cases} \frac{1}{14}x^2, & \text{if } x = 1, 2, 3, \\ 0, & \text{otherwise.} \end{cases}$$

a. The mean of X is given by

$$E[X] = \sum_{x=1,2,3} x P[X = x] = (1 \times 1/14) + (2 \times 4/14) + (3 \times 9/14) = 18/7.$$

b. The variance of X is given by

$$\begin{aligned} Var[X] &= \sum_{x=1,2,3} x^2 P[X = x] - (E[X])^2 \\ &= (1 \times 1/14) + (4 \times 4/14) + (9 \times 9/14) - (18/7)^2 \\ &= 19/49. \end{aligned}$$

2. We begin by enumerating all possible outcomes. Let the score of the roll of the first die be given by the top row of the below table, and the score of the roll of the second die be given by the leftmost column of the below table. All possible largest scores are given as follows:

© Springer Nature Switzerland AG 2020
J. Gillard, *A First Course in Statistical Inference*,
Springer Undergraduate Mathematics Series,
https://doi.org/10.1007/978-3-030-39561-2_B

	1	2	3	4	5	6
1	1	2	3	4	5	6
2	2	2	3	4	5	6
3	3	3	3	4	5	6
4	4	4	4	4	5	6
5	5	5	5	5	5	6
6	6	6	6	6	6	6

Using this table, we can deduce that the probability mass function for the random variable X is given by

x	1	2	3	4	5	6
$P[X = x]$	1/36	3/36	5/36	7/36	9/36	11/36

The mean of X is given by

$$E[X] = \sum_{x=1,2,3,4,5,6} x P[X = x] = (1 \times 1/36) + (2 \times 3/36) + \cdots + (6 \times 11/36) = 161/36.$$

3. We begin by using the property that the probabilities (as given by the probability mass function) must sum to 1. In this case

$$f(0) + f(1) + f(2) + f(3) = \alpha + \alpha(1 + \beta) + \alpha(1 + 2\beta) + \alpha(1 + 3\beta) = 1,$$

and we can thus write $\alpha = \dfrac{1}{4 + 6\beta}$.

The mean of X is given by

$$E[X] = \sum_{x=0,1,2,3} x P[X = x] = \alpha(1 + \beta) + 2\alpha(1 + 2\beta) + 3\alpha(1 + 3\beta) = 2\alpha(3 + 7\beta) = \frac{3 + 7\beta}{2 + 3\beta},$$

where we have used our already found expression for α. When $E[X] = 2$ then $4 + 6\beta = 3 + 7\beta$ giving $\beta = 1$ and $\alpha = \frac{1}{4+6} = \frac{1}{10}$.

4. Although there are six possible outcomes, we can divide them into two categories: "a six" and "not a six". We can thus use the Binomial distribution with $n = 6$ trials with probability of obtaining a six being $p = 1/6$. Let X be the discrete random variable with this Binomial distribution. Then

$$P[X = 1] = \binom{6}{1}(1/6)(5/6)^5 \approx 0.40188.$$

5. Let X be the random variable following a Binomial distribution with n trials each with probability of success p. The mean of the Binomial distribution is np and the variance is $np(1 - p)$. For the numbers given, we can write the equations

$$np = 1$$
$$np(1 - p) = 0.5$$

and solving these gives $p = 0.5$ and $n = 2$. We can find the desired probability by using the probability mass function of the Binomial distribution as follows:

$$P[X \leq 1] = P[X = 0] + P[X = 1] = \binom{2}{0}(1/2)^0(1/2)^2 + \binom{2}{1}(1/2)(1/2) = 3/4.$$

6. Suppose the random variable X of admissions follows a Poisson distribution with mean 7, that is, $X \sim Po(\lambda)$. Then using the probability mass function

$$P[X \leq 2] = P[X = 0] + P[X = 1] + P[X = 2]$$
$$= \exp(-7)\frac{7^0}{0!} + \exp(-7)\frac{7^1}{1!} + \exp(-7)\frac{7^2}{2!}$$
$$\approx 0.0296.$$

7. To find the value of k, we use the property that the integral of the probability density function over the range of possible values of x will be equal to 1, that is,

$$\int_0^1 kx(1-x)\,dx = \frac{k}{6} = 1$$

giving $k = 6$.

a. The cumulative distribution function of X is given by

$$F(x) = P[X \leq x] = \int_0^x 6x(1-x)\,dx = 3x^2 - 2x^3.$$

b. The mean is given by

$$E[X] = \int_0^1 6x^2(1-x)\,dx = \frac{1}{2}.$$

c. The variance is given by

$$Var[X] = E[X^2] - (E[X])^2 = \int_0^1 6x^3(1-x)\,dx - \left(\frac{1}{2}\right)^2 = \frac{1}{20}.$$

8. Here let X be the random variable such that $X \sim Exp(\lambda)$. As the mean is $1/4$ then $\lambda = 4$. Hence, the probability that the bulb will last for at least 1 year is

$$P[X > 1] = 1 - P[X \le 1] = 1 - F(1) = 1 - (1 - \exp(-4)) \approx 0.01832,$$

where here we have used the cumulative distribution function $F(x)$ of the Exponential random variable X.

B.2   Problems of Chap. 2

1. Since X and Y are independent Normally distributed random variables, then the random variable $2X - 3Y$ is also Normally distributed with mean

$$E[2X - 3Y] = 2E[X] - 3E[Y] = (2 \times 15) - (3 \times 9) = 3$$

and variance

$$Var[2X - 3Y] = 4Var[X] + 9Var[Y] = (4 \times 1/4) + (9 \times 8/9) = 9.$$

Hence $(2X - 3Y) \sim N[3, 9]$ and

$$P[2X - 3Y < 6] = P\left[Z < \frac{6 - 3}{3}\right] = P[Z < 1] = 0.84134,$$

by consulting Table C.1. Since $P[2X - 3Y < 6] + P[2X - 3Y \ge 6] = 1$ then $P[2X - 3Y \ge 6] = 0.15866$.

2. Let X be the random variable of woodscrew length. We are told that $X \sim N[2, 0.05]$. Using Corollary 2.1, we can deduce that $\bar{X} \sim N[2, 0.05/200]$. So

$$\begin{aligned}
P\left[\bar{X} > 2.02\right] = 1 - P\left[\bar{X} \le 2.02\right] &= 1 - P\left[Z \le \frac{2.02 - 2.00}{\sqrt{0.05/200}}\right] \\
&\approx 1 - P[Z \le 1.26] \\
&= 1 - 0.89617 = 0.10383
\end{aligned}$$

by consulting Table C.1.

3. All possible samples, with sample mean, sample variance, and median, are given in the following table:

Sample	Mean	Variance	Median	Prob
(0, 0, 0)	0	0	0	1/216
(0, 0, 1)	1/3	1/3	0	3/216
(0, 0, 2)	2/3	4/3	0	9/216
(0, 0, 3)	1	3	0	3/216
(0, 1, 1)	2/3	1/3	1	3/216
(0, 1, 2)	1	1	1	18/216
(0, 1, 3)	4/3	7/3	1	6/216
(0, 2, 2)	4/3	4/3	2	27/216
(0, 2, 3)	5/3	7/3	2	18/216
(0, 3, 3)	2	3	3	3/216
(1, 1, 1)	1	0	1	1/216
(1, 1, 2)	4/3	1/3	1	9/216
(1, 1, 3)	5/3	4/3	1	3/216
(1, 2, 2)	5/3	1/3	2	27/216
(1, 2, 3)	2	1	2	18/216
(1, 3, 3)	7/3	4/3	3	3/216
(2, 2, 2)	2	0	2	27/216
(2, 2, 3)	7/3	1/3	2	27/216
(3, 3, 3)	3	0	3	1/216
(2, 3, 3)	8/3	1/3	3	9/216

Hence, we have the following sampling distributions for the sample mean ($\bar{X}$), sample variance (S^2), and sample median (M):

$\bar{X}$	0	1/3	2/3	1	4/3	5/3	2	7/3	8/3	3
$P(\bar{X} = \bar{x})$	1/216	3/216	12/216	22/216	42/216	48/216	48/216	30/216	9/216	1/216

S^2	0	1/3	1	4/3	7/3	3
$P(S^2 = s^2)$	30/216	78/216	36/216	42/216	24/216	6/216

M	0	1	2	3
$P(M = m)$	16/216	40/216	144/216	16/216

4. An enumeration of all possible samples of size two (denoted by the random variables (X_1, X_2)) and consequently all sample means and minimums is given below:

(x_1, x_2)	$P[(X_1, X_2) = (x_1, x_2)]$	Mean	Minimum
(10, 10)	4/25	10	10
(10, 30)	4/25	20	10
(10, 40)	8/25	25	10
(30, 30)	1/25	30	30
(30, 40)	4/25	35	30
(40, 40)	4/25	40	40

The sampling distribution of the sample mean $\bar{X}$ is

$\bar{x}$	10	20	25	30	35	40
$P[\bar{X} = \bar{x}]$	4/25	4/25	8/25	1/25	4/25	4/25

with expected value $E[\bar{X}] = (10 \times 4/25) + \cdots + (40 \times 4/25) = 26$.
The sampling distribution of the sample minimum W is

w	10	30	40
$P[W = w]$	16/25	5/25	4/25

with expected value $E[W] = (10 \times 16/25) + (30 \times 5/25) + (40 \times 4/25) = 18.8$.

5. a. An enumeration of all possible samples of size two (denoted by the random variables (X_1, X_2)), and all sample means, sample variances, and sample standard deviations (to answer the later parts of the question) is given below:

(x_1, x_2)	$P[(X_1, X_2) = (x_1, x_2)]$	$\bar{x}$	s^2	s
$(1, 1)$	1/16	1	0	0
$(1, 3)$	1/8	2	2	$\sqrt{2}$
$(1, 5)$	1/8	3	8	$2\sqrt{2}$
$(1, 7)$	1/8	4	18	$3\sqrt{2}$
$(3, 3)$	1/16	3	0	0
$(3, 5)$	1/8	4	2	$\sqrt{2}$
$(3, 7)$	1/8	5	8	$2\sqrt{2}$
$(5, 5)$	1/16	5	0	0
$(5, 7)$	1/8	6	2	$\sqrt{2}$
$(7, 7)$	1/16	7	0	0

The sampling distribution of the mean $\bar{X}$ is thus

$\bar{x}$	1	2	3	4	5	6	7
$P[\bar{X} = \bar{x}]$	1/16	1/8	3/16	1/4	3/16	1/8	1/16

b. The sampling distribution of the sample variance S^2 is

s^2	0	2	8	18
$P[S^2 = s^2]$	1/4	3/8	1/4	1/8

c. Using the sampling distributions, we can compute

$$E[\bar{X}] = (1 \times 1/16) + \cdots + (7 \times 1/16) = 4,$$
$$E[S^2] = (0 \times 1/4) + \cdots + (18 \times 1/8) = 37/2,$$

and

$$Var\left[\bar{X}\right] = E\left[\bar{X}^2\right] - \left(E\left[\bar{X}\right]\right)^2 = \left(1^2 \times 1/16\right) + \cdots + \left(7^2 \times 1/16\right) - 4^2 = 5/2.$$

We know that $E\left[\bar{X}\right] = E\left[X\right] = 4$, that is, we expect the sample mean to equal the population mean. Likewise, $Var\left[\bar{X}\right] = Var[X]/n$ and so the population variance can be computed to be $2 \times 5/2 = 5$.

d. The sampling distribution of the sample standard deviation S is

and

s	0	$\sqrt{2}$	$2\sqrt{2}$	$3\sqrt{2}$
$P[S = s]$	1/4	3/8	1/4	1/8

$$E\left[S\right] = (0 \times 1/4) + \cdots + \left(3\sqrt{2} \times 1/8\right) = 5\sqrt{2}/4.$$

We can see that

$$(E\left[S\right])^2 = 25/6 \neq 5 = E\left[S^2\right].$$

6. Let S denote the lifetime of the piece of equipment. This will be the time until the first failure of a component. Hence, $S = \min(T_1, T_2, T_3, T_4, T_5)$, where T_i is the lifetime of the ith component. From Theorem 2.3, S has the probability density function

$$g(s) = 5f(s)(1 - F(s))^4,$$

where f and F denote, respectively, the probability density and cumulative density functions of T.

Here

$$F(t) = \int_0^t \frac{2}{(1+x)^3}\, dx = \left[-\frac{1}{(1+x)^2}\right]_0^t = 1 - \frac{1}{(1+t)^2}.$$

Therefore, $g(s) = 5 \times \dfrac{2}{(1+s)^3} \times \dfrac{1}{(1+s)^8} = \dfrac{10}{(1+s)^{11}}.$

Hence

$$E[S] = \int_0^\infty \frac{10s}{(1+s)^{11}}\, ds$$

$$= 10 \int_1^\infty \frac{u - 1}{u^{11}}\, du \quad \text{where } 1 + s = u$$

$$= 10 \left[-\frac{1}{9u^9} + \frac{1}{10u^{10}}\right]_1^\infty = \frac{1}{9}$$

and the mean lifetime of the piece of equipment is $\dfrac{1000}{9} = 111.1$ h.

7. Using integration by parts it can be shown that, for the distribution given, $E[X] = 1/2$ and $Var[X] = 1/4$. Alternatively, the distribution could be identified as an Exponential distribution, and then the mean and variance can be stated. Let $X_1, X_2, \ldots, X_{50}$ be a random sample of the waiting times. We are interested in the total time $T = \sum_{i=1}^{50} X_i$. By the Central Limit Theorem $T \sim N[50 \times 1/2, 50 \times 1/4] = N[25, 12.5]$ approximately. The probability that a motorist will have to wait more than 25 min before reaching the front of the queue is $P[T > 25] = P[Z > 0] = 0.5$.

We assume that the sample of 50 is large enough so that the Central Limit Theorem is a valid approximation to the true probability, and that each random variable X_i is independently and identically distributed.

B.3   Problems of Chap. 3

1. Since we expect the sample mean to equal the population mean, an unbiased estimator of the population mean is the sample mean. The sample mean is given by

$$\frac{4 + 2 + 3 + 1 + 5}{5} = \frac{15}{5} = 3.$$

We expect the sample variance to equal the population variance and so an unbiased estimator of the population variance is the sample variance which is given by

$$s^2 = \frac{1}{n-1}\left\{\sum_{i=1}^{n} x_i^2 - n\bar{x}^2\right\} = \frac{1}{4}\left\{4^2 + 2^2 + 3^2 + 1^2 + 5^2 - (5\times3^2)\right\} = 5/2.$$

2. X is the number of heads observed on tossing a coin n times, with probability of success p. It should be clear that X is a Binomial random variable, that is, $X \sim Bin(n, p)$. θ is the probability of obtaining two consecutive heads. In two independent tosses of the coin, it follows that $\theta = p \times p = p^2$. $\hat{\theta}$ is an unbiased estimator of p^2 if $E[\hat{\theta}] = p^2$. We have that

$$E[\hat{\theta}] = E\left[\frac{X(X-1)}{n(n-1)}\right] = \frac{1}{n(n-1)}\left[E[X^2] - E[X]\right].$$

Recall that X follows a Binomial distribution and $Var[X] = E[X^2] - (E[X])^2$ hence $E[X^2] = np(1-p) + n^2p^2$.
So

$$E[\hat{\theta}] = \frac{1}{n(n-1)}\left[E[X^2] - E[X]\right] = \frac{1}{n(n-1)}\left[np(1-p) + n^2p^2 - np\right] = p^2 = \theta.$$

3. Let X_1, X_2, X_3 denote the weights of the random sample of boxes. Below we list the possible random samples, corresponding sample means $(\bar{X})$, and sample medians (M).

x_1, x_2, x_3	$\bar{x}$	m	$Prob$
2,2,2	2	2	1/64
2,2,3	7/3	2	6/64
2,2,4	8/3	2	3/64
2,3,3	8/3	3	12/64
2,3,4	3	3	12/64
2,4,4	10/3	4	3/64
3,3,3	3	3	8/64
3,3,4	10/3	3	12/64
3,4,4	11/3	4	6/64
4,4,4	4	4	1/64

The sampling distributions of $\bar{X}$ and M are, respectively:

$\bar{x}$	2	7/3	8/3	3	10/3	11/3	4
$P[\bar{X} = \bar{x}]$	1/64	6/64	15/64	20/64	15/64	6/64	1/64

m	2	3	4
$P[M = m]$	10/64	44/64	10/64

Here it can be computed that $E[\bar{X}]=3$ and $E[M]=3$. It follows that $E[1000\bar{X}] = E[1000M] = 3000$.

Since the total weight of all the boxes is 3000kg, then $1000\bar{X}$ and $1000M$ are unbiased estimators of this total weight and the mean square errors of both estimators are their variances (since the bias is zero). Again from the sampling distributions we can calculate $Var[\bar{X}] = 1/6$ and $Var[M] = 20/64$. Since $Var[\bar{X}] < Var[M]$ then $Var[1000\bar{X}] < Var[1000M]$. Hence, the estimator of the total weight of all the boxes with smallest mean square error (or just smallest variance in this case) is $1000\bar{X}$.

4. Let X denote the number of stoppages. $X \sim Po(\lambda)$. There were 96 days, out of 1000, with no stoppages. We can use this information to estimate $P[X = 0]$. $\widehat{P[X = 0]} = 96/1000 = 0.096$. For a Poisson distribution $P[X = x] = \dfrac{\exp(-\lambda)\lambda^x}{x!}$ and so $P[X = 0] = \dfrac{\exp(-\lambda)\lambda^0}{0!} = \exp(-\lambda)$. Hence $\exp(-\lambda) \approx 0.096$ and solving gives $\hat{\lambda} = 2.3434$.

5. a. The cumulative distribution function of X is given by

$$F(x; \theta) = \int_0^x \frac{1}{\theta} \exp\left(-\frac{t}{\theta}\right) dt = 1 - \exp\left(-\frac{x}{\theta}\right).$$

Hence, the probability distribution function of the sample minimum W is

$$h(w; \theta) = n \times \frac{1}{\theta} \exp\left(-\frac{x}{\theta}\right) \times \left[\exp\left(-\frac{x}{\theta}\right)\right]^{n-1} = \frac{n}{\theta} \exp\left(-\frac{nx}{\theta}\right).$$

b. The probability distribution function of the sample minimum above corresponds to a probability distribution function of an Exponential distribution with mean θ/n, hence $E[W] = \theta/n$. It can thus be seen that $E[nW] = \theta$ and an unbiased estimator of θ is nW.

6. We begin by computing

$$E[T] = \int_0^\infty \alpha^2 t^2 \exp(-\alpha t)\, dt = 2/\alpha,$$

by using integration by parts. The mode is the value of t which maximizes $f(t) = \alpha^2 t \exp(-\alpha t)$. The derivative with respect to t is

$$f'(t) = \alpha^2 \exp(-\alpha t) - \alpha^3 t \exp(-\alpha t)$$

and the stationary point corresponding to a maximum can be found from the equation $f'(t) = 0$. Thus the mode is $1/\alpha$, and the expected value $2/\alpha$ is twice the mode.

The sample mean is $\bar{t} = 513.14$, and this estimates the expected value $2/\alpha$. We can then estimate α using $\hat{\alpha} = 2/513.14 \approx 0.003898$. The estimate of the mode (since the expected value is twice the mode) is 256.57.

The variance of the sample mean is given by the population variance divided by the sample size (in this case, 5). We begin by working out the variance of T.

$$E[T^2] = \int_0^\infty \alpha^2 t^3 \exp(-\alpha t)\, dt = 6/\alpha^2,$$

and hence

$$Var[T] = E[T^2] - (E[T])^2 = 6/\alpha^2 - (2/\alpha)^2 = 2/\alpha^2.$$

The variance can be estimated by $2/(0.003898)^2 \approx 131656.45$. As $n = 5$ and $Var[\bar{T}] = Var[T]/n$, the estimated variance of the sample mean is $131656.45/5 \approx 26331.29$.

Our estimate of the mode can be written as $\bar{T}/2$. So the variance of the mode is given by

$$Var\left[\bar{T}/2\right] = Var\left[\bar{T}\right]/4 \approx 26331.29/4 \approx 6582.82.$$

7. Consider the estimator $\hat{\theta}_3$. It follows that

$$E[\hat{\theta}_3] = \lambda E[\hat{\theta}_1] + (1 - \lambda)E[\hat{\theta}_2]$$
$$= \lambda\theta + (1 - \lambda)\theta$$
$$= \theta$$

and so $\hat{\theta}_3$ is an unbiased estimator of θ. We can write the variance as

$$Var[\hat{\theta}_3] = \lambda^2 Var[\hat{\theta}_1] + (1 - \lambda)^2 Var[\hat{\theta}_2]$$
$$= \lambda^2\sigma_1^2 + (1 - \lambda)^2\sigma_2^2 .$$

Differentiating $Var[\hat{\theta}_3]$ with respect to λ, we see that at a stationary point

$$2\lambda\sigma_1^2 - 2(1 - \lambda)\sigma_2^2 = 0.$$

Solving for λ we obtain $\lambda = \dfrac{\sigma_2^2}{\sigma_1^2 + \sigma_2^2}$.

Differentiating again, we obtain $2\sigma_1^2 + 2\sigma_2^2 > 0$ so the stationary point is a maximum.

a. $\bar{X}_1$ and $\bar{X}_2$ are both unbiased estimators for μ with variances $\dfrac{\sigma_1^2}{n_1}$ and $\dfrac{\sigma_2^2}{n_2}$, respectively. Using the above formula, the value of λ which minimizes the variance of $\hat{\mu}$ is thus

$$\lambda = \frac{\frac{\sigma_2^2}{n_2}}{\frac{\sigma_1^2}{n_1} + \frac{\sigma_2^2}{n_2}} = \frac{n_1\sigma_2^2}{n_2\sigma_1^2 + n_1\sigma_2^2} .$$

b. S_1^2 and S_2^2 are unbiased estimators of σ^2, with variances $\dfrac{2\sigma^4}{n_1 - 1}$ and $\dfrac{2\sigma^4}{n_2 - 1}$, respectively. In a similar manner to (a), substituting the relevant values into the equation derived at the outset of this question, the value of λ which minimizes the variance of $\hat{\sigma}^2$ is thus

$$\lambda = \frac{n_1 - 1}{n_1 + n_2 - 2} .$$

8. The probability density function of X is given by

$$f(x) = \frac{1}{\theta}, \quad 0 \le x \le \theta$$

and its cumulative distribution function is given by

$$F(x) = \frac{x}{\theta}, \quad 0 \le x \le \theta .$$

If Z is the sample maximum and $g(z)$ is its probability density function,

$$g(z) = nf(z)[F(z)]^{n-1} = n\frac{z^{n-1}}{\theta^n}, \quad 0 \le z \le \theta.$$

Thus

$$E[Z] = \int_0^\theta zn\frac{z^{n-1}}{\theta^n}\, dz = \frac{n\theta}{n+1}.$$

Hence, Z is a biased estimator of θ. However, its bias may be corrected. Consider the estimator $\frac{n+1}{n}Z$. It follows that

$$E\left[\frac{n+1}{n}Z\right] = \frac{n+1}{n}E[Z] = \frac{n+1}{n}\frac{n\theta}{n+1} = \theta$$

and so the estimator $\frac{n+1}{n}Z$ is an unbiased estimator of θ.

B.4   Problems of Chap. 4

1. a. Let μ_X denote the population mean of the first population introduced in the question, and let μ_Y denote the population mean of the second population introduced.

   The 95% confidence interval for μ_X is

   $$80 \pm 1.96 \times \sqrt{\frac{25}{25}} = [78.04, 81.96].$$

   The 95% confidence interval for μ_Y is

   $$76 \pm 1.96 \times \sqrt{\frac{25}{25}} = [74.04, 77.96].$$

   b. The 95% confidence interval for $\mu_X - \mu_Y$ is

   $$(80 - 76) \pm 1.96 \times \sqrt{\frac{25}{25} + \frac{25}{25}} = [1.2281, 6.7719].$$

2. a. For the data given $\bar{x} = 1.05$ and $s^2 = 0.015$. Assuming an independent and identically distributed sample has been taken from a Normal population, a 95% confidence interval for the population mean is

   $$\bar{x} \pm t_{0.975}(5)\sqrt{\frac{s^2}{n}} = 1.05 \pm 2.571 \times \sqrt{\frac{0.015}{6}} = [0.92145, 1.17855].$$

b. Assuming an independent and identically distributed sample has been taken from a Normal population, a 95% confidence interval for the population variance is

$$\left[\frac{(n-1)s^2}{\chi^2_{0.975}(n-1)}, \frac{(n-1)s^2}{\chi^2_{0.025}(n-1)}\right] = \left[\frac{5 \times 0.015}{12.833}, \frac{5 \times 0.015}{0.831}\right]$$

$$= [0.005844, 0.09025].$$

3. a. The 99% confidence interval for μ is

$$\bar{x} \pm 2.5758\frac{\sigma}{\sqrt{n}},$$

since $z_{0.995} = 2.5758$ and $\bar{x}$ can be seen to be located in the middle of the confidence interval. The midpoint of the confidence interval given is

$$(27.7533 + 30.1327)/2 = 28.943 = \bar{x}.$$

b. To work out the sample size n that was used in the above formula, one can use either endpoint of the confidence interval. It follows that

$$28.943 - 2.5758 \times \frac{4}{\sqrt{n}} = 27.7533.$$

Solving for n gives that $n = 75$, approximately.

4. a. Assuming the data are sampled randomly from a Normal population, the 95% confidence interval for the pop. variance of drug X is

$$\left[\frac{(m-1)s_X^2}{\chi^2_{0.975}(m-1)}, \frac{(m-1)s_X^2}{\chi^2_{0.025}(m-1)}\right] = \left[\frac{13 \times 124.8764835}{24.736}, \frac{13 \times 124.8764835}{5.009}\right]$$

$$= [65.62881, 324.09548].$$

b. The 95% confidence interval for the pop. variance of drug Y is

$$\left[\frac{(n-1)s_Y^2}{\chi^2_{0.975}(n-1)}, \frac{(n-1)s_Y^2}{\chi^2_{0.025}(n-1)}\right] = \left[\frac{9 \times 88.61}{19.023}, \frac{9 \times 88.61}{2.700}\right]$$

$$= [41.92241, 295.36667].$$

c. The confidence interval for the ratio of variances is given by

$$\left[\frac{s_X^2}{s_Y^2} \times \frac{1}{F_{0.975}(13, 9)}, \frac{s_X^2}{s_Y^2} \times \frac{1}{F_{0.025}(13, 9)}\right] = [0.36786, 4.66650]$$

which provides some evidence that the scientists claim of unequal variances may be incorrect. It could be valuable to revisit the previous analyses where unequal variances were assumed, and ascertain the difference (if any) upon assuming equal variances.

5. The sample mean and sample variance for the girl's times were 144.74 and 37.093, respectively. The sample mean and sample variance for the boy's times were 134.52 and 51.557, respectively. Assuming that both random samples are from Normal distributions with equal population variances, then we can estimate the population variance by

$$\hat{\sigma}^2 = s^2 = \frac{(5-1) \times 37.093 + (5-1) \times 51.557}{5+5-2} = 44.325.$$

The 95th percentile of Student's t-distribution with 8 degrees of freedom is 1.86, that is, $t_{0.95}(8) = 1.86$. Hence, the 90% confidence interval for the difference in population means is given by

$$(144.74 - 134.52) \pm 1.86 \times \sqrt{44.325 \left(\frac{1}{5} + \frac{1}{5} \right)} = [2.3881, 18.0519].$$

In obvious notation, the 95% confidence interval for the ratio of population variances is

$$\left[\frac{s_{girl}^2}{s_{boy}^2} \frac{1}{F_{0.975}(4, 4)}, \frac{s_{girl}^2}{s_{boy}^2} \frac{1}{F_{0.025}(4, 4)} \right] = \left[\frac{s_{girl}^2}{s_{boy}^2} \frac{1}{F_{0.975}(4, 4)}, \frac{s_{girl}^2}{s_{boy}^2} F_{0.975}(4, 4) \right]$$

$$= \left[\frac{37.093}{55.557} \times \frac{1}{9.605}, \frac{37.093}{55.557} \times 9.605 \right]$$

$$= [0.0749, 6.9104].$$

B.5   Problems of Chap. 5

1. a. The sample care 101.54 and 39.418, respectively. The sample mean and sample variance for Tiddot watches are 111.96 and 147.478, respectively.
   b. Assuming both samples are random samples from independent Normally distributed random variables with the same population variance, then

$$T = \frac{(\bar{x} - \bar{y}) - (\mu_X - \mu_Y)}{s\sqrt{\frac{1}{m} + \frac{1}{n}}} \sim t(m + n - 2)$$

where μ_X and μ_Y are the population means of Romex watches and Tiddot watches, respectively, $m = n = 5$ are the corresponding sample sizes, and $\bar{x}$ and $\bar{y}$ are the corresponding sample means. The pooled estimate of the common population variance s^2 is given by

$$s^2 = \frac{(4 \times 39.418) + (4 \times 147.478)}{8} = 93.448.$$

The null hypothesis is $H_0 : \mu_X = \mu_Y$, and the alternative hypothesis is $H_1 : \mu_X \neq \mu_Y$. Assuming H_0 is true, then the test statistic is

$$t = \frac{(101.54 - 111.96)}{\sqrt{93.448 \left(\frac{1}{5} + \frac{1}{5} \right)}} = -1.704.$$

For a 5% significance level, the acceptance region is

$$\mathcal{A} = \{t : -t_{0.975}(8) \leq t \leq t_{0.975}(8)\} = \{t : -2.306 \leq t \leq 2.306\},$$

and as our test statistic lies in $\mathcal{A}$ we accept H_0 at a 5% significance level. From tables, $t_{0.9}(8) = 1.397$ and $t_{0.95}(8) = 1.860$. As we have a two-tailed test, the p-value is given by $P[T < -1.704] + P[T > 1.704] = 2P[T > 1.704]$. Hence, the p-value for this hypothesis test is between 0.1 and 0.2.

2. Let μ_X denote the population mean mark of students studying economics. Here $H_0 : \mu_X = 40$, $H_1 : \mu_X > 40$. For the given data $\bar{x} = 51.42857$ and $s_X^2 = 677.2857$. The test statistic is

$$t = \frac{51.42857 - 40}{\sqrt{\frac{677.2857}{7}}} = 1.16186.$$

For a 5% significance level, the acceptance region is

$$\mathcal{A} = \{t : -t_{0.975}(6) \leq t \leq t_{0.975}(6)\} = \{t : -2.447 \leq t \leq 2.447\},$$

and as our test statistic is in $\mathcal{A}$ we accept H_0 at a 5% significance level.
The assumptions are that the data is independently and identically Normally distributed.
The second sample is assumed to be independent of the first, and the Normal populations from which both samples are taken are assumed to have equal variances. Let μ_Y denote the population mean mark after the introduction of a student support service. For the given data $\bar{y} = 55.42857$ and $s_Y^2 = 1214.952$. The pooled variance estimator is

$$s^2 = \frac{(6 \times 677.2857) + (6 \times 1214.952)}{12} = 946.119.$$

The test statistic under the null hypothesis H_0 is

$$t = \frac{55.42857 - 51.42857}{\sqrt{946.119(\frac{1}{7} + \frac{1}{7})}} = 0.243288.$$

For a 5% significance level the acceptance region is

$$\mathcal{A} = \{t : -t_{0.975}(12) \leq t \leq t_{0.975}(12)\} = \{t : -2.179 \leq t \leq 2.179\},$$

and as our test statistic is in $\mathcal{A}$ we accept H_0 at a 5% significance level.

3. Let p be the probability of success. The hypotheses are $H_0 : p = 0.6$ and $H_1 : p \neq 0.6$. An estimator of p is $\hat{p} = 106/200 = 0.66$. Using the Central Limit Theorem and assuming H_0 is true, the test statistic is

$$z = \frac{0.66 - 0.6}{\sqrt{\frac{0.6 \times 0.4}{200}}} = 1.73205 .$$

For a 5% significance level the acceptance region is

$$\mathcal{A} = \{z : -z_{0.975} \leq z \leq z_{0.975}\} = \{z : -1.96 \leq t \leq 1.96\},$$

and as our test statistic is in $\mathcal{A}$ we accept H_0 at a 5% significance level.

4. a. For the data given, in obvious notation, $s_{South}^2 = 88.455560$ and $s_{North}^2 = 50.4$. The hypotheses are $H_0 : \sigma_{South}^2 = \sigma_{North}^2$ and $H_1 : \sigma_{South}^2 \neq \sigma_{North}^2$. Under the null hypothesis H_0, a suitable test statistic can be computed as

$$f = \frac{s_{South}^2/\sigma_{South}^2}{s_{North}^2/\sigma_{North}^2} = \frac{s_{South}^2}{s_{North}^2} = \frac{88.455560}{50.4} \approx 1.76 .$$

For a 5% significance level the acceptance region $\mathcal{A}$ is given by

$$\mathcal{A} = \{f : 0 \leq f \leq F_{1-\alpha}(m - 1, n - 1)\} = \{f : 0 \leq f \leq F_{0.95}(9, 9)\}$$
$$= \{f : 0 \leq f \leq 3.179\} .$$

The test statistic f is in the acceptance region $\mathcal{A}$ and so we accept the null hypothesis at a 5% significance level.

b. Using the result of the previous question, we proceed with assuming that both population variances σ_{South}^2 and σ_{North}^2 are equal.

We estimate the common population variance by using the pooled variance estimator

$$s^2 = \frac{(9 \times 88.455560) + (9 \times 50.4)}{18} = 69.427778 .$$

The hypotheses are $H_0 : \mu_{South} = \mu_{North}$ and $H_1 : \mu_{South} \neq \mu_{North}$. The sample means are $\bar{x}_{South} = 28.7$ and $\bar{x}_{North} = 23.2$. Under the null hypothesis, an appropriate test statistic is

$$t = \frac{28.7 - 23.2}{\sqrt{69.427778 \left(\frac{1}{10} + \frac{1}{10}\right)}} \approx 1.47598 .$$

For a 5% significance level the acceptance region is

$$\mathcal{A} = \{t : -t_{0.975}(18) \leq t \leq t_{0.975}(18)\} = \{t : -2.101 \leq t \leq 2.101\} ,$$

and our value of the test statistic t lies in this region. We hence accept the null hypothesis that the population means are equal at a 5% significance level.

5. In the table below we list the expected frequencies should a Poisson distribution with mean $\mu = 1.0$ model the number of ships arriving. If X is the random variable of the number of ships arriving, then

$$P[X = x] = \frac{\exp(-\mu)\mu^x}{x!},$$

and for example

$$P[X = 0] = \frac{\exp(-1.0) \times 1.0^0}{0!} = 0.3679.$$

Hence, the expected frequency corresponding to 0 ships arriving in 70 h is $0.3679 \times 70 = 25.753$. This exercise can be completed for all remaining entries bar the last category "3 or more" whose expected frequency can be found by noting that all the expected frequencies should sum to 70.

No. of ships arriving	Frequency	Prob	Exp. Frequency
0	42	0.3679	25.753
1	21	0.3679	25.753
2	7	0.1839	12.873
3 or more	0	0.0803	5.621

The test statistic is given by $\chi^2 = 19.44$. The acceptance region is

$$\mathcal{A} = \{\chi^2 : 0 \leq \chi^2 \leq \chi^2_{0.95}(3)\} = \{\chi^2 : 0 \leq \chi^2 \leq 7.815\},$$

and the null hypothesis is rejected at a 5% significance level.

6. The critical region is given by $C = \{\bar{x} : |\bar{x}| > 1.96/\sqrt{n}\}$ and the power function $\pi(\mu) = P[C; \mu] = P[|\bar{X}| > 1.96/\sqrt{n}; \mu]$.
   Since $\bar{X} \sim N[\mu, 1/n]$, the power function can be written as

$$\pi(\mu) = 1 - P\left[-1.96 - \sqrt{n}\mu \leq Z \leq 1.96 - \sqrt{n}\mu\right].$$

It is possible to evaluate this power function for various μ. I have done this using a computer but you would not have to compute all of these values to have enough information to sketch the power function.
For example, below are pairs $(\mu, \pi(\mu))$ for $n = 25$:

$$(\pm 0.5, 0.705), (\pm 0.4, 0.516), (\pm 0.3, 0.323), (\pm 0.2, 0.170), (\pm 0.1, 0.079), (0, 0.05).$$

The figure below is a plot of the power function for $n = 25$ (red, dotted line), $n = 100$ (blue, dashed line), and $n = 500$ (black, solid line). The hypothesis test becomes more "powerful" as the sample size increases.

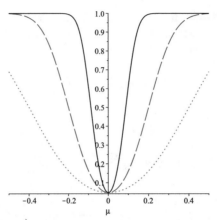

7. The power function is given by

$$\pi(\theta) = P[\mathcal{C}; \theta] = P[Z \leq 0.5; \theta] \,.$$

To obtain an expression for this probability, we need to find the cumulative density function of the sample maximum Z. The probability density function $f(x)$ and cumulative density function $F(x)$ of X are

$$f(x) = 1/\theta, \qquad F(x) = \int_0^x 1/\theta \, dx = x/\theta,$$

for $0 \leq x \leq \theta$.

The cumulative distribution function $G(z)$ for the sample maximum Z is given by (using Theorem 2.3)

$$G(z) = [F(z)]^n = [z/\theta]^5 = z^5/\theta^5 \,.$$

So $\pi(\theta) = P(Z \leq 0.5; \theta) = G(0.5) = 0.5^5/\theta^5 = 0.03125/\theta^5$. The significance level can be determined by evaluating the power function at the parameter specified in the null hypothesis, giving $\alpha = \pi(1) = 0.03125$.

B.6   Problems of Chap. 6

1. The ANOVA table is given by

Source of variation	df	SS	MS	F-ratio
Between groups	3	521000.000	173666.667	4.354
Within groups	8	319066.667	39883.333	
Total	11	840066.667		

From statistical tables $F_{0.95}(3, 8) = 4.066$ and so we reject the hypothesis of equal population means at a 5% significance level.

By Fisher's least significant difference test, the kth and lth groups will have population means that are significantly different if the difference of the sample means is bigger than

$$t_{0.975}(8)\sqrt{39883.333 \times \frac{2}{3}} = 2.306 \times \sqrt{39883.333 \times \frac{2}{3}} = 376.02.$$

The mean of each variety is (in increasing order) $\begin{matrix} 1 & 4 & 2 & 3 \\ 10 & 53.333 & 100 & 530 \end{matrix}$ and so variety 3 has significantly different population means from all other varieties.

2. The ANOVA table is given by

Source of variation	df	SS	MS	F-ratio
Between groups	3	1.474333	0.491444	95.27194
Within groups	8	0.041267	0.005158	
Total	11	1.5156		

From statistical tables $F_{0.95}(3, 8) = 4.066$ and so we reject the hypothesis of equal population means at a 5% significance level.

For a 5% significance level, and as the group sizes are equal, Fisher's LSD criterion is

$$t_{0.975}(8)\sqrt{s^2\left(\frac{1}{3} + \frac{1}{3}\right)} = 2.306\sqrt{0.005158\left(\frac{1}{3} + \frac{1}{3}\right)} = 0.13551,$$

and in summary we observe the following:

Group	Group	Difference in means	Conclusion
1	2	−0.8567	Different
1	3	−0.1867	Different
1	4	−0.01	Not different
2	3	0.67	Different
2	4	0.8467	Different
3	4	0.1767	Different

3. a. For the total sum of squares,

$$SS_{total} = \sum_{i=1}^{m}\sum_{j=1}^{n_i}(x_{ij} - \bar{x}_{\bullet\bullet})^2$$

$$= \sum_{i=1}^{m}\sum_{j=1}^{n_i}(x_{ij}^2 - 2\bar{x}_{\bullet\bullet}x_{ij} + \bar{x}_{\bullet\bullet}^2)$$

$$= \sum_{i=1}^{m}\sum_{j=1}^{n_i}x_{ij}^2 - 2\bar{x}_{\bullet\bullet}\sum_{i=1}^{m}\sum_{j=1}^{n_i}x_{ij} + \sum_{i=1}^{m}\sum_{j=1}^{n_i}\bar{x}_{\bullet\bullet}^2$$

$$= \sum_{i=1}^{m} \sum_{j=1}^{n_i} x_{ij}^2 - 2\bar{x}_{\bullet\bullet} x_{\bullet\bullet} + (n_1 x_{\bullet\bullet}^2 + n_2 x_{\bullet\bullet}^2 + \cdots + n_m x_{\bullet\bullet}^2)$$

$$= \sum_{i=1}^{m} \sum_{j=1}^{n_i} x_{ij}^2 - \frac{2x_{\bullet\bullet}^2}{N} + Nx_{\bullet\bullet}^2$$

$$= \sum_{i=1}^{m} \sum_{j=1}^{n_i} x_{ij}^2 - \frac{2x_{\bullet\bullet}^2}{N} + \frac{x_{\bullet\bullet}^2}{N}$$

$$= \sum_{i=1}^{m} \sum_{j=1}^{n_i} x_{ij}^2 - \frac{x_{\bullet\bullet}^2}{N}$$

b. For the between-group sum of squares,

$$SS_G = \sum_{i=1}^{m} \sum_{j=1}^{n_i} (\bar{x}_{i\bullet} - \bar{x}_{\bullet\bullet})^2$$

$$= \sum_{i=1}^{m} n_i (\bar{x}_{i\bullet} - \bar{x}_{\bullet\bullet})^2$$

$$= \sum_{i=1}^{m} n_i \bar{x}_{i\bullet}^2 - 2\bar{x}_{\bullet\bullet} \sum_{i=1}^{m} n_i \bar{x}_{i\bullet} + \bar{x}_{\bullet\bullet} \sum_{i=1}^{m} n_i$$

$$= \sum_{i=1}^{m} n_i \left(\frac{x_{i\bullet}}{n_i}\right)^2 - 2\bar{x}_{\bullet\bullet} \sum_{i=1}^{m} x_{i\bullet} + N\bar{x}_{\bullet\bullet}^2$$

$$= \sum_{i=1}^{m} \frac{x_{i\bullet}^2}{n_i} - 2N\bar{x}_{\bullet\bullet}^2 + N\bar{x}_{\bullet\bullet}^2$$

$$= \sum_{i=1}^{m} \frac{x_{i\bullet}^2}{n_i} - N\bar{x}_{\bullet\bullet}^2 = \sum_{i=1}^{m} \frac{x_{i\bullet}^2}{n_i} - \frac{x_{\bullet\bullet}^2}{N}$$

c. For the within-group sum of squares, consider

$$\sum_{j=1}^{n_i} (x_{ij} - \bar{x}_{i\bullet})^2 = \sum_{j=1}^{n_i} x_{ij}^2 - 2\bar{x}_{i\bullet} x_{ij} + \bar{x}_{i\bullet}^2$$

$$= \sum_{j=1}^{n_i} x_{ij}^2 - 2n_i \bar{x}_{i\bullet}^2 + n_i \bar{x}_{i\bullet}^2$$

$$= \sum_{j=1}^{n_i} x_{ij}^2 - n_i \bar{x}_{i\bullet}^2$$

$$= \sum_{j=1}^{n_i} x_{ij}^2 - \frac{x_{i\bullet}^2}{n_i}.$$

Hence

$$SS_{error} = \sum_{i=1}^{m} \sum_{j=1}^{n_i} (x_{ij} - \bar{x}_{i\bullet})^2$$

$$= \sum_{i=1}^{m} \left(\sum_{j=1}^{n_i} x_{ij}^2 - \frac{x_{i\bullet}^2}{n_i} \right)$$

$$= \sum_{i=1}^{m} \sum_{j=1}^{n_i} x_{ij}^2 - \sum_{i=1}^{m} \frac{x_{i\bullet}^2}{n_i}$$

B.7   Problems of Chap. 7

1. Let x denote "age" and Y denote "cost". From the data $\bar{x} = 3$, $\bar{y} = 2.5$, $\sum_{i=1}^{n} x_i^2 = 80$, $\sum_{i=1}^{n} y_i^2 = 39.5$, $\sum_{i=1}^{n} x_i y_i = 51.5$. Hence, the slope of the regression line is estimated as

$$\hat{\beta} = \frac{S_{xy}}{S_{xx}} = \frac{\sum_{i=1}^{n} x_i y_i - n\bar{x}\bar{y}}{\sum_{i=1}^{n} x_i^2 - n\bar{x}^2} = \frac{51.5 - (6 \times 3 \times 2.5)}{60 - 6 \times 3^2} = 0.25,$$

and the intercept is estimated as $\hat{\alpha} = \bar{y} - \hat{\beta}\bar{x} = 2.5 - 0.25 \times 3 = 1.75$. When $x = 2$, $\hat{y} = 1.75 + 0.25 \times 2 = 2.25$.

2. Note that $\bar{x} = 25$, $\bar{y} = 17.66$, $S_{xx} = 10$, $S_{yy} = 7.052$, $S_{xy} = 8.1$. We can thus compute that $\hat{\alpha} = -2.59$ and $\hat{\beta} = 0.81$.

   We need to make some distributional assumptions concerning the random error term ε_i of the model:

   a. They are all independent.
   b. They have zero expectation.
   c. They all have the same variance σ^2.
   d. They are Normally distributed.

   Then random variable

$$T = \frac{\hat{\beta} - \beta}{\sqrt{\frac{s^2}{S_{xx}}}}$$

   has the Student's t-distribution with $n - 2$ degrees of freedom.

   We have that $SS_{error} = S_{yy} - \frac{S_{xy}^2}{S_{xx}} = 7.052 - \frac{8.1^2}{10} = 0.491$ and so $s^2 =$

$0.491/3 = 0.16367$. Under the null hypothesis that $\beta = 0$ then

$$t = \frac{0.81 - 0}{\sqrt{\frac{0.16367}{10}}} = 6.3315.$$

The acceptance region for the test is

$$A = \{t : -t_{0.995}(3) \leq t \leq t_{0.995}(3)\}.$$

From statistical tables $t_{0.995}(3) = 5.841$ and so we reject the null hypothesis that $\beta = 0$ at a 1% significance level.

3. Using the formulas $\hat{\alpha} = \bar{y} - \hat{\beta}\bar{x}$ and $\hat{\beta} = \dfrac{S_{xy}}{S_{xx}}$, we obtain that $\hat{\alpha} = -1.640$ and $\hat{\beta} = 0.773$. Note that $S_{xx} = 82.5$, $S_{xy} = 63.75$, and $S_{yy} = 49.9290$. Also we calculate $s^2 = 0.083455$.

To test the null hypothesis that the slope of the line relating x and y is equal to zero, $H_0 : \beta = 0$, $H_1 : \beta \neq 0$, it follows that

$$T = \frac{\hat{\beta} - \beta}{\sqrt{\frac{s^2}{S_{xx}}}} \sim t(n - 2)$$

and assuming H_0 true

$$t = \frac{0.773 - 0}{\sqrt{\frac{0.0830}{82.5}}} = 24.2956.$$

Note that we make the same assumptions as listed for the previous question. The acceptance region for the test is

$$A = \{t : -t_{0.995}(8) \leq t \leq t_{0.995}(8)\}.$$

From statistical tables $t_{0.995}(8) = 3.355$ and so we reject the null hypothesis that $\beta = 0$ at a 1% significance level.
Using the formula

$$\hat{y} \pm t_{1-\alpha/2}(n - 2)\sqrt{s^2\left(\frac{1}{n} + \frac{(x_0 - \bar{x})^2}{S_{xx}}\right)}$$

with $\hat{y} = \hat{\alpha} + \hat{\beta} \times x_0 = 16.133$, $x_0 = 23$ and $t_{0.975}(8) = 2.306$, the 95% confidence interval for the head girth of a fish of length 23 cm is given by [15.741, 16.524].

Statistical Tables

C

C.1   The Normal Distribution Function

© Springer Nature Switzerland AG 2020

J. Gillard, *A First Course in Statistical Inference*,

Springer Undergraduate Mathematics Series,

https://doi.org/10.1007/978-3-030-39561-2_C

Table C.1 This table gives the probability p that a Normally distributed random variable Z with zero mean and unit variance is less than or equal to z

z	0.00	0.01	0.02	0.03	0.04	0.05	0.06	0.07	0.08	0.09
0.0	0.50000	0.50399	0.50798	0.51197	0.51595	0.51994	0.52392	0.52790	0.53188	0.53586
0.1	0.53983	0.54380	0.54776	0.55172	0.55567	0.55962	0.56356	0.56749	0.57142	0.57535
0.2	0.57926	0.58317	0.58706	0.59095	0.59483	0.59871	0.60257	0.60642	0.61026	0.61409
0.3	0.61791	0.62172	0.62552	0.62930	0.63307	0.63683	0.64058	0.64431	0.64803	0.65173
0.4	0.65542	0.65910	0.66276	0.66640	0.67003	0.67364	0.67724	0.68082	0.68439	0.68793
0.5	0.69146	0.69497	0.69847	0.70194	0.70540	0.70884	0.71226	0.71566	0.71904	0.72240
0.6	0.72575	0.72907	0.73237	0.73565	0.73891	0.74215	0.74537	0.74857	0.75175	0.75490
0.7	0.75804	0.76115	0.76424	0.76730	0.77035	0.77337	0.77637	0.77935	0.78230	0.78524
0.8	0.78814	0.79103	0.79389	0.79673	0.79955	0.80234	0.80511	0.80785	0.81057	0.81327
0.9	0.81594	0.81859	0.82121	0.82381	0.82639	0.82894	0.83147	0.83398	0.83646	0.83891
1.0	0.84134	0.84375	0.84614	0.84849	0.85083	0.85314	0.85543	0.85769	0.85993	0.86214
1.1	0.86433	0.86650	0.86864	0.87076	0.87286	0.87493	0.87698	0.87900	0.88100	0.88298
1.2	0.88493	0.88686	0.88877	0.89065	0.89251	0.89435	0.89617	0.89796	0.89973	0.90147
1.3	0.90320	0.90490	0.90658	0.90824	0.90988	0.91149	0.91309	0.91466	0.91621	0.91774
1.4	0.91924	0.92073	0.92220	0.92364	0.92507	0.92647	0.92785	0.92922	0.93056	0.93189
1.5	0.93319	0.93448	0.93574	0.93699	0.93822	0.93943	0.94062	0.94179	0.94295	0.94408

(continued)

Table C.1 (continued)

	0.00	0.01	0.02	0.03	0.04	0.05	0.06	0.07	0.08	0.09
1.6	0.94520	0.94630	0.94738	0.94845	0.94950	0.95053	0.95154	0.95254	0.95352	0.95449
1.7	0.95543	0.95637	0.95728	0.95818	0.95907	0.95994	0.96080	0.96164	0.96246	0.96327
1.8	0.96407	0.96485	0.96562	0.96638	0.96712	0.96784	0.96856	0.96926	0.96995	0.97062
1.9	0.97128	0.97193	0.97257	0.97320	0.97381	0.97441	0.97500	0.97558	0.97615	0.97670
2.0	0.97725	0.97778	0.97831	0.97882	0.97932	0.97982	0.98030	0.98077	0.98124	0.98169
2.1	0.98214	0.98257	0.98300	0.98341	0.98382	0.98422	0.98461	0.98500	0.98537	0.98574
2.2	0.98610	0.98645	0.98679	0.98713	0.98745	0.98778	0.98809	0.98840	0.98870	0.98899
2.3	0.98928	0.98956	0.98983	0.99010	0.99036	0.99061	0.99086	0.99111	0.99134	0.99158
2.4	0.99180	0.99202	0.99224	0.99245	0.99266	0.99286	0.99305	0.99324	0.99343	0.99361
2.5	0.99379	0.99396	0.99413	0.99430	0.99446	0.99461	0.99477	0.99492	0.99506	0.99520
2.6	0.99534	0.99547	0.99560	0.99573	0.99585	0.99598	0.99609	0.99621	0.99632	0.99643
2.7	0.99653	0.99664	0.99674	0.99683	0.99693	0.99702	0.99711	0.99720	0.99728	0.99736
2.8	0.99744	0.99752	0.99760	0.99767	0.99774	0.99781	0.99788	0.99795	0.99801	0.99807
2.9	0.99813	0.99819	0.99825	0.99831	0.99836	0.99841	0.99846	0.99851	0.99856	0.99861
3.0	0.99865	0.99869	0.99874	0.99878	0.99882	0.99886	0.99889	0.99893	0.99896	0.99900
3.1	0.99903	0.99906	0.99910	0.99913	0.99916	0.99918	0.99921	0.99924	0.99926	0.99929
3.2	0.99931	0.99934	0.99936	0.99938	0.99940	0.99942	0.99944	0.99946	0.99948	0.99950
3.3	0.99952	0.99953	0.99955	0.99957	0.99958	0.99960	0.99961	0.99962	0.99964	0.99965
3.4	0.99966	0.99968	0.99969	0.99970	0.99971	0.99972	0.99973	0.99974	0.99975	0.99976
3.5	0.99977	0.99978	0.99978	0.99979	0.99980	0.99981	0.99981	0.99982	0.99983	0.99983
3.6	0.99984	0.99985	0.99985	0.99986	0.99986	0.99987	0.99987	0.99988	0.99988	0.99989
3.7	0.99989	0.99990	0.99990	0.99990	0.99991	0.99991	0.99992	0.99992	0.99992	0.99992
3.8	0.99993	0.99993	0.99993	0.99994	0.99994	0.99994	0.99994	0.99995	0.99995	0.99995
3.9	0.99995	0.99995	0.99996	0.99996	0.99996	0.99996	0.99996	0.99996	0.99997	0.99997

C.2   Percentage Points of the Normal Distribution

Table C.2 This table gives the values of z satisfying $P[Z \leq z] = p$ where Z is a Normally distributed random variable with zero mean and unit variance

p	0	0.01	0.02	0.03	0.04	0.05	0.06	0.07	0.08	0.09
0.50	0.0000	0.0251	0.0502	0.0753	0.1004	0.1257	0.1510	0.1764	0.2019	0.2275
0.60	0.2533	0.2793	0.3055	0.3319	0.3585	0.3853	0.4125	0.4399	0.4677	0.4959
0.70	0.5244	0.5534	0.5828	0.6128	0.6433	0.6745	0.7063	0.7388	0.7722	0.8064
0.80	0.8416	0.8779	0.9154	0.9542	0.9945	1.0364	1.0803	1.1264	1.1750	1.2265
0.90	1.2816	1.3408	1.4051	1.4758	1.5548	1.6449	1.7507	1.8808	2.0537	2.3263
0.95	1.6449	1.6546	1.6646	1.6747	1.6849	1.6954	1.7060	1.7169	1.7279	1.7392
0.96	1.7507	1.7624	1.7744	1.7866	1.7991	1.8119	1.8250	1.8384	1.8522	1.8663
0.97	1.8808	1.8957	1.9110	1.9268	1.9431	1.9600	1.9774	1.9954	2.0141	2.0335
0.98	2.0537	2.0749	2.0969	2.1201	2.1444	2.1701	2.1973	2.2262	2.2571	2.2904
0.99	2.3263	2.3656	2.4089	2.4573	2.5121	2.5758	2.6521	2.7478	2.8782	3.0902

C.3   Percentage Points of the Chi-Squared Distribution

Table C.3 Values of k satisfying $P[X \leq k] = p$ where X is a χ^2 random variable with ν degrees of freedom

ν	p									
	0.005	0.01	0.025	0.05	0.1	0.9	0.95	0.975	0.99	0.995
1	0.000	0.000	0.001	0.004	0.016	2.706	3.841	5.024	6.635	7.879
2	0.010	0.020	0.051	0.103	0.211	4.605	5.991	7.378	9.210	10.597
3	0.072	0.115	0.216	0.352	0.584	6.251	7.815	9.348	11.345	12.838
4	0.207	0.297	0.484	0.711	1.064	7.779	9.488	11.143	13.277	14.860
5	0.412	0.554	0.831	1.145	1.610	9.236	11.070	12.833	15.086	16.750
6	0.676	0.872	1.237	1.635	2.204	10.645	12.592	14.449	16.812	18.548
7	0.989	1.239	1.690	2.167	2.833	12.017	14.067	16.013	18.475	20.278

Table C.3 (continued)

v	p									
	0.005	0.01	0.025	0.05	0.1	0.9	0.95	0.975	0.99	0.995
8	1.344	1.646	2.180	2.733	3.490	13.362	15.507	17.535	20.090	21.955
9	1.735	2.088	2.700	3.325	4.168	14.684	16.919	19.023	21.666	23.589
10	2.156	2.558	3.247	3.940	4.865	15.987	18.307	20.483	23.209	25.188
11	2.603	3.053	3.816	4.575	5.578	17.275	19.675	21.920	24.725	26.757
12	3.074	3.571	4.404	5.226	6.304	18.549	21.026	23.337	26.217	28.300
13	3.565	4.107	5.009	5.892	7.042	19.812	22.362	24.736	27.688	29.819
14	4.075	4.660	5.629	6.571	7.790	21.064	23.685	26.119	29.141	31.319
15	4.601	5.229	6.262	7.261	8.547	22.307	24.996	27.488	30.578	32.801
16	5.142	5.812	6.908	7.962	9.312	23.542	26.296	28.845	32.000	34.267
17	5.697	6.408	7.564	8.672	10.085	24.769	27.587	30.191	33.409	35.718
18	6.265	7.015	8.231	9.390	10.865	25.989	28.869	31.526	34.805	37.156
19	6.844	7.633	8.907	10.117	11.651	27.204	30.144	32.852	36.191	38.582
20	7.434	8.260	9.591	10.851	12.443	28.412	31.410	34.170	37.566	39.997
21	8.034	8.897	10.283	11.591	13.240	29.615	32.671	35.479	38.932	41.401
22	8.643	9.542	10.982	12.338	14.041	30.813	33.924	36.781	40.289	42.796
23	9.260	10.196	11.689	13.091	14.848	32.007	35.172	38.076	41.638	44.181
24	9.886	10.856	12.401	13.848	15.659	33.196	36.415	39.364	42.980	45.559
25	10.520	11.524	13.120	14.611	16.473	34.382	37.652	40.646	44.314	46.928
26	11.160	12.198	13.844	15.379	17.292	35.563	38.885	41.923	45.642	48.290
27	11.808	12.879	14.573	16.151	18.114	36.741	40.113	43.195	46.963	49.645
28	12.461	13.565	15.308	16.928	18.939	37.916	41.337	44.461	48.278	50.993
29	13.121	14.256	16.047	17.708	19.768	39.087	42.557	45.722	49.588	52.336
30	13.787	14.953	16.791	18.493	20.599	40.256	43.773	46.979	50.892	53.672
31	14.458	15.655	17.539	19.281	21.434	41.422	44.985	48.232	52.191	55.003
32	15.134	16.362	18.291	20.072	22.271	42.585	46.194	49.480	53.486	56.328
33	15.815	17.074	19.047	20.867	23.110	43.745	47.400	50.725	54.776	57.648
34	16.501	17.789	19.806	21.664	23.952	44.903	48.602	51.966	56.061	58.964
35	17.192	18.509	20.569	22.465	24.797	46.059	49.802	53.203	57.342	60.275
36	17.887	19.233	21.336	23.269	25.643	47.212	50.998	54.437	58.619	61.581
37	18.586	19.960	22.106	24.075	26.492	48.363	52.192	55.668	59.893	62.883
38	19.289	20.691	22.878	24.884	27.343	49.513	53.384	56.896	61.162	64.181
39	19.996	21.426	23.654	25.695	28.196	50.660	54.572	58.120	62.428	65.476
40	20.707	22.164	24.433	26.509	29.051	51.805	55.758	59.342	63.691	66.766
45	24.311	25.901	28.366	30.612	33.350	57.505	61.656	65.410	69.957	73.166
50	27.991	29.707	32.357	34.764	37.689	63.167	67.505	71.420	76.154	79.490
55	31.735	33.570	36.398	38.958	42.060	68.796	73.311	77.380	82.292	85.749

C.4   Percentage Points of the Student's t-Distribution

Table C.4 Values of k satisfying $P[X \leq k] = p$ where X is a random variable having the Student's t-distribution with v degrees of freedom

v	p				
	0.9	0.95	0.975	0.99	0.995
1	3.078	6.314	12.706	31.821	63.657
2	1.886	2.920	4.303	6.965	9.925
3	1.638	2.353	3.182	4.541	5.841
4	1.533	2.132	2.776	3.747	4.604
5	1.476	2.015	2.571	3.365	4.032
6	1.440	1.943	2.447	3.143	3.707
7	1.415	1.895	2.365	2.998	3.499
8	1.397	1.860	2.306	2.896	3.355
9	1.383	1.833	2.262	2.821	3.250
10	1.372	1.812	2.228	2.764	3.169
11	1.363	1.796	2.201	2.718	3.106
12	1.356	1.782	2.179	2.681	3.055
13	1.350	1.771	2.160	2.650	3.012
14	1.345	1.761	2.145	2.624	2.977
15	1.341	1.753	2.131	2.602	2.947
16	1.337	1.746	2.120	2.583	2.921
17	1.333	1.740	2.110	2.567	2.898
18	1.330	1.734	2.101	2.552	2.878
19	1.328	1.729	2.093	2.539	2.861
20	1.325	1.725	2.086	2.528	2.845
21	1.323	1.721	2.080	2.518	2.831
22	1.321	1.717	2.074	2.508	2.819
23	1.319	1.714	2.069	2.500	2.807
24	1.318	1.711	2.064	2.492	2.797
25	1.316	1.708	2.060	2.485	2.787
26	1.315	1.706	2.056	2.479	2.779
27	1.314	1.703	2.052	2.473	2.771
28	1.313	1.701	2.048	2.467	2.763
29	1.311	1.699	2.045	2.462	2.756

Table C.4 (continued)

ν	p				
	0.9	0.95	0.975	0.99	0.995
30	1.310	1.697	2.042	2.457	2.750
31	1.309	1.696	2.040	2.453	2.744
32	1.309	1.694	2.037	2.449	2.738
33	1.308	1.692	2.035	2.445	2.733
34	1.307	1.691	2.032	2.441	2.728
35	1.306	1.690	2.030	2.438	2.724
36	1.306	1.688	2.028	2.434	2.719
37	1.305	1.687	2.026	2.431	2.715
38	1.304	1.686	2.024	2.429	2.712
39	1.304	1.685	2.023	2.426	2.708
40	1.303	1.684	2.021	2.423	2.704
45	1.301	1.679	2.014	2.412	2.690
50	1.299	1.676	2.009	2.403	2.678
55	1.297	1.673	2.004	2.396	2.668
60	1.296	1.671	2.000	2.390	2.660

C.5   Percentage Points of the F-Distribution

Values of k satisfying $P[X \leq k] = p$ where X is the random variable having the F-distribution with ν_1 and ν_2 degrees of freedom. To find percentage points in the lower tail use

$$F_p(\nu_1, \nu_2) = \frac{1}{F_{1-p}(\nu_2, \nu_1)}$$

Table C.5 The table below corresponds to $p = 0.995$

v_2 \ v_1	1	2	3	4	5	6	7	8	9	10	11	12	15	20	25	30	40	50	100
1	16211	20000	21615	22500	23056	23437	23715	23925	24091	24224	24334	24426	24630	24836	24960	25044	25148	25211	25337
2	198.501	199.000	199.166	199.250	199.300	199.333	199.357	199.375	199.388	199.400	199.409	199.416	199.433	199.450	199.460	199.466	199.475	199.480	199.490
3	55.552	49.799	47.467	46.195	45.392	44.838	44.434	44.126	43.882	43.686	43.524	43.387	43.085	42.778	42.591	42.466	42.308	42.213	42.022
4	31.333	26.284	24.259	23.155	22.456	21.975	21.622	21.352	21.139	20.967	20.824	20.705	20.438	20.167	20.002	19.892	19.752	19.667	19.497
5	22.785	18.314	16.530	15.556	14.940	14.513	14.200	13.961	13.772	13.618	13.491	13.384	13.146	12.903	12.755	12.656	12.530	12.454	12.300
6	18.635	14.544	12.917	12.028	11.464	11.073	10.786	10.566	10.391	10.250	10.133	10.034	9.814	9.589	9.451	9.358	9.241	9.170	9.026
7	16.236	12.404	10.882	10.050	9.522	9.155	8.885	8.678	8.514	8.380	8.270	8.176	7.968	7.754	7.623	7.534	7.422	7.354	7.217
8	14.688	11.042	9.596	8.805	8.302	7.952	7.694	7.496	7.339	7.211	7.104	7.015	6.814	6.608	6.482	6.396	6.288	6.222	6.088
9	13.614	10.107	8.717	7.956	7.471	7.134	6.885	6.693	6.541	6.417	6.314	6.227	6.032	5.832	5.708	5.625	5.519	5.454	5.322
10	12.826	9.427	8.081	7.343	6.872	6.545	6.302	6.116	5.968	5.847	5.746	5.661	5.471	5.274	5.153	5.071	4.966	4.902	4.772
11	12.226	8.912	7.600	6.881	6.422	6.102	5.865	5.682	5.537	5.418	5.320	5.236	5.049	4.855	4.736	4.654	4.551	4.488	4.359
12	11.754	8.510	7.226	6.521	6.071	5.757	5.525	5.345	5.202	5.085	4.988	4.906	4.721	4.530	4.412	4.331	4.228	4.165	4.037
13	11.374	8.186	6.926	6.233	5.791	5.482	5.253	5.076	4.935	4.820	4.724	4.643	4.460	4.270	4.153	4.073	3.970	3.908	3.780
14	11.060	7.922	6.680	5.998	5.562	5.257	5.031	4.857	4.717	4.603	4.508	4.428	4.247	4.059	3.942	3.862	3.760	3.698	3.569
15	10.798	7.701	6.476	5.803	5.372	5.071	4.847	4.674	4.536	4.424	4.329	4.250	4.070	3.883	3.766	3.687	3.585	3.523	3.394
16	10.575	7.514	6.303	5.638	5.212	4.913	4.692	4.521	4.384	4.272	4.179	4.099	3.920	3.734	3.618	3.539	3.437	3.375	3.246
17	10.384	7.354	6.156	5.497	5.075	4.779	4.559	4.389	4.254	4.142	4.050	3.971	3.793	3.607	3.492	3.412	3.311	3.248	3.119
18	10.218	7.215	6.028	5.375	4.956	4.663	4.445	4.276	4.141	4.030	3.938	3.860	3.683	3.498	3.382	3.303	3.201	3.139	3.009
19	10.073	7.093	5.916	5.268	4.853	4.561	4.345	4.177	4.043	3.933	3.841	3.763	3.587	3.402	3.287	3.208	3.106	3.043	2.913
20	9.944	6.986	5.818	5.174	4.762	4.472	4.257	4.090	3.956	3.847	3.756	3.678	3.502	3.318	3.203	3.123	3.022	2.959	2.828
25	9.475	6.598	5.462	4.835	4.433	4.150	3.939	3.776	3.645	3.537	3.447	3.370	3.196	3.013	2.898	2.819	2.716	2.652	2.519
30	9.180	6.355	5.239	4.623	4.228	3.949	3.742	3.580	3.450	3.344	3.255	3.179	3.006	2.823	2.708	2.628	2.524	2.459	2.323

(continued)

Table C.5 (continued)

v_2	v_1																			
	1	2	3	4	5	6	7	8	9	10	11	12	15	20	25	30	40	50	100
40	8.828	6.066	4.976	4.374	3.986	3.713	3.509	3.350	3.222	3.117	3.028	2.953	2.781	2.598	2.482	2.401	2.296	2.230	2.088
50	8.626	5.902	4.826	4.232	3.849	3.579	3.376	3.219	3.092	2.988	2.900	2.825	2.653	2.470	2.353	2.272	2.164	2.097	1.951
100	8.241	5.589	4.542	3.963	3.589	3.325	3.127	2.972	2.847	2.744	2.657	2.583	2.411	2.227	2.108	2.024	1.912	1.840	1.681

Table C.6 The table below corresponds to $p = 0.99$

v_2 \ v_1	1	2	3	4	5	6	7	8	9	10	11	12	15	20	25	30	40	50	100
1	4052	5000	5403	5625	5764	5859	5928	5981	6022	6056	6083	6106	6157	6209	6240	6261	6287	6303	6334
2	98.503	99.000	99.166	99.249	99.299	99.333	99.356	99.374	99.388	99.399	99.408	99.416	99.433	99.449	99.459	99.466	99.474	99.479	99.489
3	34.116	30.817	29.457	28.710	28.237	27.911	27.672	27.489	27.345	27.229	27.133	27.052	26.872	26.690	26.579	26.505	26.411	26.354	26.240
4	21.198	18.000	16.694	15.977	15.522	15.207	14.976	14.799	14.659	14.546	14.452	14.374	14.198	14.020	13.911	13.838	13.745	13.690	13.577
5	16.258	13.274	12.060	11.392	10.967	10.672	10.456	10.289	10.158	10.051	9.963	9.888	9.722	9.553	9.449	9.379	9.291	9.238	9.130
6	13.745	10.925	9.780	9.148	8.746	8.466	8.260	8.102	7.976	7.874	7.790	7.718	7.559	7.396	7.296	7.229	7.143	7.091	6.987
7	12.246	9.547	8.451	7.847	7.460	7.191	6.993	6.840	6.719	6.620	6.538	6.469	6.314	6.155	6.058	5.992	5.908	5.858	5.755
8	11.259	8.649	7.591	7.006	6.632	6.371	6.178	6.029	5.911	5.814	5.734	5.667	5.515	5.359	5.263	5.198	5.116	5.065	4.963
9	10.561	8.022	6.992	6.422	6.057	5.802	5.613	5.467	5.351	5.257	5.178	5.111	4.962	4.808	4.713	4.649	4.567	4.517	4.415
10	10.044	7.559	6.552	5.994	5.636	5.386	5.200	5.057	4.942	4.849	4.772	4.706	4.558	4.405	4.311	4.247	4.165	4.115	4.014
11	9.646	7.206	6.217	5.668	5.316	5.069	4.886	4.744	4.632	4.539	4.462	4.397	4.251	4.099	4.005	3.941	3.860	3.810	3.708
12	9.330	6.927	5.953	5.412	5.064	4.821	4.640	4.499	4.388	4.296	4.220	4.155	4.010	3.858	3.765	3.701	3.619	3.569	3.467
13	9.074	6.701	5.739	5.205	4.862	4.620	4.441	4.302	4.191	4.100	4.025	3.960	3.815	3.665	3.571	3.507	3.425	3.375	3.272
14	8.862	6.515	5.564	5.035	4.695	4.456	4.278	4.140	4.030	3.939	3.864	3.800	3.656	3.505	3.412	3.348	3.266	3.215	3.112
15	8.683	6.359	5.417	4.893	4.556	4.318	4.142	4.004	3.895	3.805	3.730	3.666	3.522	3.372	3.278	3.214	3.132	3.081	2.977
16	8.531	6.226	5.292	4.773	4.437	4.202	4.026	3.890	3.780	3.691	3.616	3.553	3.409	3.259	3.165	3.101	3.018	2.967	2.863
17	8.400	6.112	5.185	4.669	4.336	4.102	3.927	3.791	3.682	3.593	3.519	3.455	3.312	3.162	3.068	3.003	2.920	2.869	2.764
18	8.285	6.013	5.092	4.579	4.248	4.015	3.841	3.705	3.597	3.508	3.434	3.371	3.227	3.077	2.983	2.919	2.835	2.784	2.678
19	8.185	5.926	5.010	4.500	4.171	3.939	3.765	3.631	3.523	3.434	3.360	3.297	3.153	3.003	2.909	2.844	2.761	2.709	2.602
20	8.096	5.849	4.938	4.431	4.103	3.871	3.699	3.564	3.457	3.368	3.294	3.231	3.088	2.938	2.843	2.778	2.695	2.643	2.535
25	7.770	5.568	4.675	4.177	3.855	3.627	3.457	3.324	3.217	3.129	3.056	2.993	2.850	2.699	2.604	2.538	2.453	2.400	2.289
30	7.562	5.390	4.510	4.018	3.699	3.473	3.304	3.173	3.067	2.979	2.906	2.843	2.700	2.549	2.453	2.386	2.299	2.245	2.131

(continued)

Table C.6 (continued)

ν_2	ν_1																			
	1	2	3	4	5	6	7	8	9	10	11	12	15	20	25	30	40	50	100
40	7.314	5.179	4.313	3.828	3.514	3.291	3.124	2.993	2.888	2.801	2.727	2.665	2.522	2.369	2.271	2.203	2.114	2.058	1.938
50	7.171	5.057	4.199	3.720	3.408	3.186	3.020	2.890	2.785	2.698	2.625	2.562	2.419	2.265	2.167	2.098	2.007	1.949	1.825
100	6.895	4.824	3.984	3.513	3.206	2.988	2.823	2.694	2.590	2.503	2.430	2.368	2.223	2.067	1.965	1.893	1.797	1.735	1.598

Table C.7 The table below corresponds to $p = 0.975$

ν_2 \ ν_1	1	2	3	4	5	6	7	8	9	10	11	12	15	20	25	30	40	50	100
1	648	800	864	900	922	937	948	957	963	969	973	977	985	993	998	1001	1006	1009	1013
2	38.506	39.000	39.165	39.248	39.298	39.331	39.355	39.373	39.387	39.398	39.407	39.415	39.431	39.448	39.458	39.465	39.473	39.478	39.488
3	17.443	16.044	15.439	15.101	14.885	14.735	14.624	14.540	14.473	14.419	14.374	14.337	14.253	14.167	14.115	14.081	14.037	14.010	13.956
4	12.218	10.649	9.979	9.605	9.364	9.197	9.074	8.980	8.905	8.844	8.794	8.751	8.657	8.560	8.501	8.461	8.411	8.381	8.319
5	10.007	8.434	7.764	7.388	7.146	6.978	6.853	6.757	6.681	6.619	6.568	6.525	6.428	6.329	6.268	6.227	6.175	6.144	6.080
6	8.813	7.260	6.599	6.227	5.988	5.820	5.695	5.600	5.523	5.461	5.410	5.366	5.269	5.168	5.107	5.065	5.012	4.980	4.915
7	8.073	6.542	5.890	5.523	5.285	5.119	4.995	4.899	4.823	4.761	4.709	4.666	4.568	4.467	4.405	4.362	4.309	4.276	4.210
8	7.571	6.059	5.416	5.053	4.817	4.652	4.529	4.433	4.357	4.295	4.243	4.200	4.101	3.999	3.937	3.894	3.840	3.807	3.739
9	7.209	5.715	5.078	4.718	4.484	4.320	4.197	4.102	4.026	3.964	3.912	3.868	3.769	3.667	3.604	3.560	3.505	3.472	3.403
10	6.937	5.456	4.826	4.468	4.236	4.072	3.950	3.855	3.779	3.717	3.665	3.621	3.522	3.419	3.355	3.311	3.255	3.221	3.152
11	6.724	5.256	4.630	4.275	4.044	3.881	3.759	3.664	3.588	3.526	3.474	3.430	3.330	3.226	3.162	3.118	3.061	3.027	2.956
12	6.554	5.096	4.474	4.121	3.891	3.728	3.607	3.512	3.436	3.374	3.321	3.277	3.177	3.073	3.008	2.963	2.906	2.871	2.800
13	6.414	4.965	4.347	3.996	3.767	3.604	3.483	3.388	3.312	3.250	3.197	3.153	3.053	2.948	2.882	2.837	2.780	2.744	2.671
14	6.298	4.857	4.242	3.892	3.663	3.501	3.380	3.285	3.209	3.147	3.095	3.050	2.949	2.844	2.778	2.732	2.674	2.638	2.565
15	6.200	4.765	4.153	3.804	3.576	3.415	3.293	3.199	3.123	3.060	3.008	2.963	2.862	2.756	2.689	2.644	2.585	2.549	2.474
16	6.115	4.687	4.077	3.729	3.502	3.341	3.219	3.125	3.049	2.986	2.934	2.889	2.788	2.681	2.614	2.568	2.509	2.472	2.396
17	6.042	4.619	4.011	3.665	3.438	3.277	3.156	3.061	2.985	2.922	2.870	2.825	2.723	2.616	2.548	2.502	2.442	2.405	2.329
18	5.978	4.560	3.954	3.608	3.382	3.221	3.100	3.005	2.929	2.866	2.814	2.769	2.667	2.559	2.491	2.445	2.384	2.347	2.269
19	5.922	4.508	3.903	3.559	3.333	3.172	3.051	2.956	2.880	2.817	2.765	2.720	2.617	2.509	2.441	2.394	2.333	2.295	2.217
20	5.871	4.461	3.859	3.515	3.289	3.128	3.007	2.913	2.837	2.774	2.721	2.676	2.573	2.464	2.396	2.349	2.287	2.249	2.170
25	5.686	4.291	3.694	3.353	3.129	2.969	2.848	2.753	2.677	2.613	2.560	2.515	2.411	2.300	2.230	2.182	2.118	2.079	1.996

(continued)

Table C.7 (continued)

ν_2	ν_1																		
	1	2	3	4	5	6	7	8	9	10	11	12	15	20	25	30	40	50	100
30	5.568	4.182	3.589	3.250	3.026	2.867	2.746	2.651	2.575	2.511	2.458	2.412	2.307	2.195	2.124	2.074	2.009	1.968	1.882
40	5.424	4.051	3.463	3.126	2.904	2.744	2.624	2.529	2.452	2.388	2.334	2.288	2.182	2.068	1.994	1.943	1.875	1.832	1.741
50	5.340	3.975	3.390	3.054	2.833	2.674	2.553	2.458	2.381	2.317	2.263	2.216	2.109	1.993	1.919	1.866	1.796	1.752	1.656
100	5.179	3.828	3.250	2.917	2.696	2.537	2.417	2.321	2.244	2.179	2.124	2.077	1.968	1.849	1.770	1.715	1.640	1.592	1.483

Table C.8 The table below corresponds to $p = 0.95$

v_2 \ v_1	1	2	3	4	5	6	7	8	9	10	11	12	15	20	25	30	40	50	100
1	161	200	216	225	230	234	237	239	241	242	243	244	246	248	249	250	251	252	253
2	18.513	19.000	19.164	19.247	19.296	19.330	19.353	19.371	19.385	19.396	19.405	19.413	19.429	19.446	19.456	19.462	19.471	19.476	19.486
3	10.128	9.552	9.277	9.117	9.013	8.941	8.887	8.845	8.812	8.786	8.763	8.745	8.703	8.660	8.634	8.617	8.594	8.581	8.554
4	7.709	6.944	6.591	6.388	6.256	6.163	6.094	6.041	5.999	5.964	5.936	5.912	5.858	5.803	5.769	5.746	5.717	5.699	5.664
5	6.608	5.786	5.409	5.192	5.050	4.950	4.876	4.818	4.772	4.735	4.704	4.678	4.619	4.558	4.521	4.496	4.464	4.444	4.405
6	5.987	5.143	4.757	4.534	4.387	4.284	4.207	4.147	4.099	4.060	4.027	4.000	3.938	3.874	3.835	3.808	3.774	3.754	3.712
7	5.591	4.737	4.347	4.120	3.972	3.866	3.787	3.726	3.677	3.637	3.603	3.575	3.511	3.445	3.404	3.376	3.340	3.319	3.275
8	5.318	4.459	4.066	3.838	3.687	3.581	3.500	3.438	3.388	3.347	3.313	3.284	3.218	3.150	3.108	3.079	3.043	3.020	2.975
9	5.117	4.256	3.863	3.633	3.482	3.374	3.293	3.230	3.179	3.137	3.102	3.073	3.006	2.936	2.893	2.864	2.826	2.803	2.756
10	4.965	4.103	3.708	3.478	3.326	3.217	3.135	3.072	3.020	2.978	2.943	2.913	2.845	2.774	2.730	2.700	2.661	2.637	2.588
11	4.844	3.982	3.587	3.357	3.204	3.095	3.012	2.948	2.896	2.854	2.818	2.788	2.719	2.646	2.601	2.570	2.531	2.507	2.457
12	4.747	3.885	3.490	3.259	3.106	2.996	2.913	2.849	2.796	2.753	2.717	2.687	2.617	2.544	2.498	2.466	2.426	2.401	2.350
13	4.667	3.806	3.411	3.179	3.025	2.915	2.832	2.767	2.714	2.671	2.635	2.604	2.533	2.459	2.412	2.380	2.339	2.314	2.261
14	4.600	3.739	3.344	3.112	2.958	2.848	2.764	2.699	2.646	2.602	2.565	2.534	2.463	2.388	2.341	2.308	2.266	2.241	2.187
15	4.543	3.682	3.287	3.056	2.901	2.790	2.707	2.641	2.588	2.544	2.507	2.475	2.403	2.328	2.280	2.247	2.204	2.178	2.123
16	4.494	3.634	3.239	3.007	2.852	2.741	2.657	2.591	2.538	2.494	2.456	2.425	2.352	2.276	2.227	2.194	2.151	2.124	2.068
17	4.451	3.592	3.197	2.965	2.810	2.699	2.614	2.548	2.494	2.450	2.413	2.381	2.308	2.230	2.181	2.148	2.104	2.077	2.020
18	4.414	3.555	3.160	2.928	2.773	2.661	2.577	2.510	2.456	2.412	2.374	2.342	2.269	2.191	2.141	2.107	2.063	2.035	1.978
19	4.381	3.522	3.127	2.895	2.740	2.628	2.544	2.477	2.423	2.378	2.340	2.308	2.234	2.155	2.106	2.071	2.026	1.999	1.940
20	4.351	3.493	3.098	2.866	2.711	2.599	2.514	2.447	2.393	2.348	2.310	2.278	2.203	2.124	2.074	2.039	1.994	1.966	1.907
25	4.242	3.385	2.991	2.759	2.603	2.490	2.405	2.337	2.282	2.236	2.198	2.165	2.089	2.007	1.955	1.919	1.872	1.842	1.779
30	4.171	3.316	2.922	2.690	2.534	2.421	2.334	2.266	2.211	2.165	2.126	2.092	2.015	1.932	1.878	1.841	1.792	1.761	1.695
40	4.085	3.232	2.839	2.606	2.449	2.336	2.249	2.180	2.124	2.077	2.038	2.003	1.924	1.839	1.783	1.744	1.693	1.660	1.589

(continued)

Table C.8 (continued)

v_2	v_1 1	2	3	4	5	6	7	8	9	10	11	12	15	20	25	30	40	50	100
50	4.034	3.183	2.790	2.557	2.400	2.286	2.199	2.130	2.073	2.026	1.986	1.952	1.871	1.784	1.727	1.687	1.634	1.599	1.525
100	3.936	3.087	2.696	2.463	2.305	2.191	2.103	2.032	1.975	1.927	1.886	1.850	1.768	1.676	1.616	1.573	1.515	1.477	1.392

Index

© Springer Nature Switzerland AG 2020
J. Gillard, *A First Course in Statistical Inference*,
Springer Undergraduate Mathematics Series,
https://doi.org/10.1007/978-3-030-39561-2_C